Radiological Issues for Fukushima's Revitalized Future

Tomoyuki Takahashi

Editor

Radiological Issues for Fukushima's Revitalized Future

Editor
Tomoyuki Takahashi
Research Reactor Institute
Kyoto University
Kumatori, Osaka, Japan

ISBN 978-4-431-55847-7   ISBN 978-4-431-55848-4   (eBook)
DOI 10.1007/978-4-431-55848-4

Library of Congress Control Number: 2015958094

Springer Tokyo Heidelberg New York Dordrecht London

Printed on acid-free paper

Springer Japan KK is part of Springer Science+Business Media (www.springer.com)

Foreword

When the Fukushima Daiichi nuclear power plant accident occurred in March 2011, Japanese society lost trust in nuclear safety. Many ordinary people hated radiation and radioactive materials, although quantum beams and radioisotopes are very useful for basic research and industry. We struggled to decide upon what efforts we should make in this situation, and finally we at the Kyoto University Research Reactor Institute (KURRI) started a new program called "KUR Research Program for the Scientific Basis of Nuclear Safety" in 2012.

As an educational part of this program, we started the "Across-the-Board Nuclear Educational Program". The weakness in the education about nuclear engineering is the fragmentation of different areas of expertise, although all experts know that nuclear engineering should be an integrated discipline. Therefore, we tried to take a new approach to cultivate highly educated nuclear engineers with a wide perspective and a synthesis of engineering judgment. Our institute has a wide variety of nuclear research facilities, such as research reactors (KUR and KUCA), accelerators (FFAG, cyclotron, e-linac), hot laboratory and tracer lab. We can create practical training courses for graduate students to experience the usage of radiation and/or radioisotopes. This experience will give them deeper knowledge than working on paper.

As a research module, we started the "Research Program for Scientific Basis of Nuclear Safety," which includes integrated nuclear safety research and the amassing/verification of accident data. One of the most important activities is the development of KURAMA (Kyoto University Radiation Mapping system), which is a GPS-aided mobile radiation monitoring system. It was developed in KURRI and has played an important role in evaluating the air dose rate in Fukushima and other parts of Eastern Japan.

Four international symposia also have been held annually for summarizing the results of this program. The titles of the symposia were "Environmental Monitoring and Dose Estimation of Residents After the Accident of TEPCO's Fukushima Daiichi Nuclear Power Stations" (December 14, 2012), "Nuclear Back-end and Transmutation Technology for Waste Disposal" (November 28, 2013) and "Earthquake, Tsunami and Nuclear Risks After the Accident of TEPCO's Fukushima Daiichi Nuclear Power Stations" (October 30, 2014). The final one was also held in Fukushima (May 30–31, 2015), entitled "Radiological Issues for Fukushima's Revitalized Future". It is very significant for us that the final symposium was held at the starting point of our program.

On behalf of KURRI, I wish to thank all the participants in our four symposia and the collaborators in our program. We also express our special gratitude to the people in Iizaka, Fukushima. KURRI hopes that this publication will promote further progress in nuclear safety research and will contribute to the revitalization of Fukushima.

Kyoto University Research Reactor Institute   Yuji Kawabata
Kumatori, Osaka, Japan

Preface

Four years have passed since the accident at Tokyo Electric Power Company's Fukushima Daiichi nuclear power station, and gradual steps have been taken toward environmental recovery in Fukushima. However, there are still many issues that need to be tackled in order to achieve the full revitalization of Fukushima. These issues encompass many different disciplines such as economics, psychology, and sociology. I believe the role of the sciences related to radiation and radioactivity is especially important.

An international symposium titled "Radiological Issues for Fukushima's Revitalized Future" was held in Iizaka-onsen, Fukushima, Japan, May 30–31, 2015. We invited experts, researchers, and the general public to attend this symposium and think together, through lectures, poster presentations, and panel discussions, about the concrete steps that need to be taken toward the revitalization of Fukushima.

In this symposium, 14 invited lectures and more than 70 scientific posters were presented. We selected 20 presentations for publication in this book. These manuscripts were peer-reviewed by experts in the field to form the chapters of the book and were divided into the four contents as follows:

Part 1 Radioactivity in the Terrestrial Environment

This part consists of six chapters related to the identification and migration of radionuclides in the terrestrial environment. It is important to clearly define the behavior of radionuclides, such as radiocesium in the terrestrial environment, to evaluate the long-term effect of the accident. These chapters give useful information on the present status and characteristics of radionuclides in the terrestrial environment.

Part 2 Decontamination and Radioactive Waste

This part consists of three chapters related to the safety of the decontamination system and the treatment of radioactive waste. The safe treatment and disposal of waste generated by decontamination are the most important problems needing quick resolution for the revitalization of Fukushima. These chapters focus on the safe treatment and reduction of cesium in waste.

Part 3 Environmental Radiation and External Exposure

This part consists of four chapters related to the development of a measurement system for environmental radiation and the evaluation of external exposure. For the residents of Fukushima, external exposure comes predominantly from contact with radiocesium. Therefore, development of a high-accuracy system for measuring environmental radiation and a method for evaluating external exposure are required. These chapters are a response to the request of government bodies and the public for such developments.

Part 4 Radioactivity in Foods and Internal Exposure

This part consists of seven chapters related to the identification of radionuclides in farm products, control of radionuclide root uptake, the decrease in radionuclide concentration by food processing, and the evaluation of internal exposure. Some of the chapters report that the internal dose by ingestion is not very high. However, the level of uneasiness among the public with regard to contamination of food is high. In addition, some agricultural products, such as peaches, are special products of Fukushima. These chapters provide information about the safety of food products and should help ease some of the worries of the public. This information can also be useful for the revitalization of some industries.

This book focuses on Fukushima's revitalized future. It is difficult to cover all the radiological issues; however, I believe that the advanced research presented here provides useful information and data to help in that revitalization.

Osaka, Japan   Tomoyuki Takahashi

Cooperators

Hirofumi Tsukada (Vice-Director, Institute of Environmental Radioactivity, Fukushima University)
Hirokuni Yamanishi (Kinki Univ.)
Hiromi Yamazawa (Nagoya Univ.)
Itsumasa Urabe (Fukuyama Univ.)
Noriyuki Momoshima (Kyushu Univ.)
Takeshi Iimoto (Univ. Tokyo)
Sumi Yokoyama (Fujita Health Univ.)
Minoru Yoneda (Kyoto Univ.)
Jiro Inaba (REA)
Kimiaki Saito (JAEA)
Shigeo Uchida (NIRS)
Takatoshi Hattori (CRIEPI)
Masahiro Osako (NIES)
Hajimu Yamana (NDF)
Yuji Kawabata (Director, Research Reactor Institute, Kyoto University)
Sentaro Takahashi (Vice-Director, Research Reactor Institute, Kyoto University)
Tomoyuki Takahashi (Research Reactor Institute, Kyoto University)
Satoshi Fukutani (Research Reactor Institute, Kyoto University)
Hiroshi Yashima (Research Reactor Institute, Kyoto University)
Nobuhiro Sato (Research Reactor Institute, Kyoto University)
Maki Nakatani (Research Reactor Institute, Kyoto University)

Contents

Part I
Radioactivity in the Terrestrial Environment

Chapter 1
Nuclear Magnetic Resonance Study of Cs Adsorption onto Clay Minerals

Yomei Tokuda, Yutaro Norikawa, Hirokazu Masai, Yoshikatsu Ueda, Naoto Nihei, Shigeto Fujimura, and Yuji Ono

Abstract The release of radioactive cesium into the environment in the aftermath of disasters such as the Fukushima Daiichi disaster poses a great health risk, particularly since cesium easily spreads in nature. In this context, we perform solid-state nuclear magnetic resonance (NMR) experiments to study Cs^+ ions adsorbed by clay minerals to analyze their local structure. The NMR spectra show two kinds of peaks corresponding to the clays (illite and kaolinite) after immersion in CsCl aqueous solution; the peak at -30 ppm is assigned to Cs^+ on the clay surface while that at -100 ppm is assigned to Cs^+ in the silicate sheet in the clay crystal. This result is consistent with the fact that Cs^+ with smaller coordination number yields a small field shift in the NMR spectra. Moreover, after immersion in KCl aqueous solution, these peaks disappear in the NMR spectra, thereby indicating that our assignment is reasonable. This is because Cs^+ on the clay surface and in the silicate sheet is easily subject to ion exchange by K^+. We believe that our findings will contribute to a better understanding of the pathway through which Cs transfers from the soil to plants and also to the recovery of the agriculture in Fukushima.

Y. Tokuda (✉) • Y. Norikawa • H. Masai
Institute for Chemical Research, Kyoto University, Gokasho, Uji, Kyoto 611-0011, Japan
e-mail: tokuda@noncry.kuicr.kyoto-u.ac.jp

Y. Ueda
Research Institute for Sustainable Humanosphere, Kyoto University, Gokasho, Uji, Kyoto 611-0011, Japan

N. Nihei
Graduate School of Agricultural and Life Sciences, The University of Tokyo, 1-1-1, Yayoi, Bunkyo-ku, Tokyo, Japan

S. Fujimura
NARO Tohoku Agricultural Research Center, 50 Harajukuminami, Arai, Fukushima-shi, Fukushima 960-2156, Japan

Y. Ono
Fukushima Agricultural Technology Centre, 116, Shimonakamichi, Takakura, Hiwadamachi, Koriyama, Fukushima 963-0531, Japan

© The Author(s) 2016
T. Takahashi (ed.), *Radiological Issues for Fukushima's Revitalized Future*,
DOI 10.1007/978-4-431-55848-4_1

Keywords Radioactive cesium • Cs adsorption by soil • Cs NMR spectra
• Kaolinite • Illite

1.1   Introduction

The occurrence of the Tohoku earthquake on March 11, 2011, led to the meltdown of the Fukushima Daiichi Nuclear Power Station in Japan. The accident released several kinds of radioactive elements such as ^{90}Sr, ^{134}Cs, ^{137}Cs, and ^{131}I into the environment. Human exposure to ^{137}Cs is a health risk because of its long half-life [1]. The element ^{137}Cs is mostly stabilized in the soil while small quantities are absorbed by plants such as rice [2]. In this manner, ^{137}Cs seeps into the food chain, leading to its internal exposure in humans and animals. In order to avoid internal exposure, it is important to understand the mechanism underlying its transfer from the soil to plants. Recently, it has been reported that the Cs transfer coefficient exhibits variation even for plants grown in soils with similar levels of radioactivity [3, 4]. Moreover, artificially added ^{137}Cs can be more easily absorbed by plants than stable Cs in the soil [5]. Thus, we have considered that one of the reasons for this phenomenon is the varied ways in which Cs (i.e., different Cs states) is adsorbed onto clay minerals in the soils.

The clay minerals that stabilize Cs include 1:1-type layer silicates and 2:1-type layer silicates [6]. These silicates stabilize Cs on the surface and silicate sheets because of their negative charges. In particular, in the case of 2:1-type layer silicates, Cs is strongly adsorbed at frayed edge sites (FESs) [7–10]. In this context, it is necessary to understand the stabilization mechanism to analyze the structure of Cs in the clay surface, silicate sheet, and FESs.

The concentration of Cs in the environment is of the order of parts per billion (ppb) or parts per trillion (ppt). Thus, very fine measurement techniques are required to measure such minute quantities and obtain their structural information. The technique of X-ray absorption fine structure (XAFS) spectroscopy is suitable for such measurements [11, 12]. However, the technique of XAFS requires synchrotron radiation, and hence, the method cannot be used in the laboratory. Therefore, another complementary method is required.

In this context, solid-state nuclear magnetic resonance (NMR) has been used to analyze the local structure of Cs in crystals and conventional glasses [13]. As regards the NMR spectra of cesium silicate crystals, Cs$^+$ ions with large coordination numbers (CNs) such as Cs$_6$Si$_{10}$O$_{23}$ exhibit a large field shift while those with smaller CNs (Cs$_2$Si$_2$O$_5$) exhibit a small field shift. In addition, the same relationship also holds for Cs present in mixed alkali silicate glasses. This relationship can be used to study the local structure of Cs in clay minerals. Moreover, solid-state NMR of clay minerals can be utilized to distinguish Cs on the surface and within silicate sheets [14–17]. In this study, we discuss the structure of Cs adsorbed onto two kinds of clay minerals (kaolinite as 1:1-type layer silicates and illite as 1:2-type layer silicates) by using solid state NMR together with XAFS spectroscopy.

1.2 Experimental

1.2.1 Sample Preparation

Kaolinite (Wako Chemicals), illite (G-O networks), ^{133}CsCl (Wako Chemicals), KCl (Wako Chemicals), and ultrapure water (Wako Chemicals) were used as received in our experiments. First, 5 g of illite was immersed in 50 mL of 0.01 M CsCl aqueous solution over time periods of 1 day, 1 month, 6 months, and 2 years. After immersion, the illite samples were separated by centrifugation. Next, 50 mL of ultrapure water was added to each illite sample followed by centrifugal separation. This washing process was performed twice. After washing, the illites were dried overnight at 40 °C. We referred to the various samples as illite_1d, illite_1m, illite_6m, and illite_2y depending on their immersion periods of 1 day, 1 month, 6 months, and 2 years, respectively. For comparison purposes, kaolinite was also immersed in 50 mL of 0.01 M CsCl aqueous solution over 1 day, 1 month, and 6 months. The kaolinites were also washed using the abovementioned washing process. Following the nomenclature used for the illite samples, we named the kaolinite samples as kaolinite_1d, kaolinite_1m, and kaolinite_6m.

In order to remove the Cs adsorbed onto illite, 1 g of the sample illites_2y was immersed in 50 mL of 0.01 KCl aqueous solution for 2 h and 2 days. These illites were also washed as per the abovementioned washing process. These "re-ion-exchanged" samples were referred to as illite_2y_KCl2h and illite_2y_KCl2d.

Further, pristine samples of illite and kaolinite were also analyzed. Table 1.1 lists all the analyzed samples along with the corresponding experimental conditions.

Table 1.1 Sample notations of clay minerals used in this study

Notation	Period of immersion in CsCl(aq)	Period of immersion in KCl(aq) after immersion in CsCl(aq)
illite_prisitine	–	–
illite_1d	1 day	–
illite_6m	6 months	–
illite_2y	2 years	–
illite_2y_KCl2h	2 years	2 h
illite_2y_KCl2d	2 years	2 days
kaolinite_pristine	–	–
kaolinite_1d	1 day	–
kaolinite_6m	6 months	–

1.3   Structure Analyses

The crystal structures were analyzed by powder X-ray diffraction (XRD) (RINT 2100, RIGAKU). We used a Cu X-ray source that was operated at 40 kV and 40 mA via the conventional $2\theta/\theta$ method. The diffractions were acquired at intervals of $0.02°$. The extended X-ray absorption fine structure (EXAFS) spectra were obtained at the cesium K-absorption edge via the fluorescence method (BL14B2, SPring-8). The cumulated number of measurements was 40. The XAFS spectra were analyzed by using ATHENA [18].

The solid-state ^{133}Cs NMR spectra of all the samples were acquired using a Chemagnetics CMX400 spectrometer utilizing a commercial probe (4 mm). The rotation speed was set to 10 kHz with an accuracy ± 10 Hz. At an external field of 9.4 T, the resonance frequency was set to about 103.7 MHz. For each measurement, the widths of the 90° pulses were set to 2.2 μs. The spectra were obtained with a cycle time of 10 s. The chemical shift reference was 1 mol/L CsCl aqueous solution, whose chemical shifts were set to 0 ppm.

1.4   Results

Figure 1.1 shows the XRD patterns of the illite_pristine, illite_6m, kaolinite_pristine, and kaolinite_6m samples. In the case of illite, the peak around 27° shows a shift to a higher angle after immersion in CsCl, thereby indicating a decrease in the lattice constant. On the other hand, for kaolinite, the peak around 27° shifts to a lower angle, which indicates an increase in the lattice constant.

Figure 1.2 shows k^2-weighted K-edge EXAFS spectra for the illite_2y and illite_2y_KCl2h samples. As previously reported for the radial distribution functions (RDFs), the first, second, and third peaks can be assigned to Cs–O, Cs–Si, and Cs–Cs, respectively [19, 20]. However, we have not obtained the RDF in the present stage. We can just note that there is a little change in EXAFS spectra for the illite_2y and illite_2y_KCl2h samples.

Figure 1.3 shows the ^{133}Cs NMR spectra of the illite and kaolinite samples after immersion in CsCl solution. The NMR spectra for all the illites exhibit peaks at around -30 and -100 ppm. On the other hand, the NMR spectra for kaolinite_1d and _1m exhibit only one peak at around -30 ppm. The kaolinite_6m sample (which was immersed for a longer time) exhibits a clear peak at around -30 ppm and a small peak at around -100 ppm.

The effect of re-ion-exchanging by K^+ on the NMR spectra was also studied (Fig. 1.4). The NMR peak for illite_2y_KCl2d vanished, although the peak at around -30 ppm was still observed for illite_2y_KCl2h.

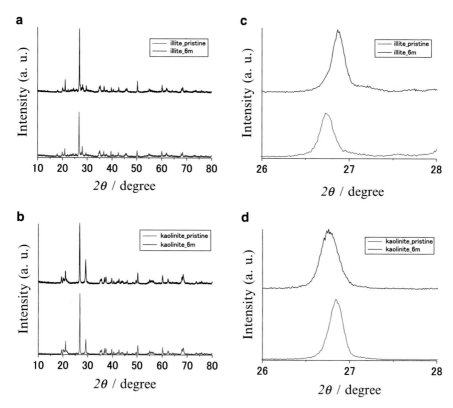

Fig. 1.1 XRD patterns of illite (**a**) as received and after immersion in KCl solution, (**b**) kaolinite as received and after immersion in KCl solution. Highest peaks around 27° assigned to (0 0 6) are also shown as (**c**) and (**d**)

1.5 Discussion

After sample immersion in CsCl solution, the lattice constant change for illite is different from that for kaolinite, as shown in Fig. 1.1; the lattice constant of illite decreases, while that of kaolinite increases. This is because hydrated K^+ in the silicate sheet is replaced by Cs^+ in illite; on the other hand, in the case of kaolinite, a proton is replaced by Cs^+. Alteration in the k^2-weighted K-edge EXAFS spectra for illite after immersion in CsCl solution as shown in Fig. 1.2 may support these observations.

As shown in Fig. 1.3, the NMR spectra of illite exhibit two peaks at -30 and -100 ppm, while those of kaolinite exhibit two peaks (one clear peak and one very small peak at -30 and -100 ppm, respectively). Kaolinite has a negative surface charge on the crystallite. Accordingly, the clear peak at -30 ppm can be assigned to the surface Cs^+. In contrast, illite has a negative charge between silicate sheets. As a result, the peaks at -100 ppm can be assigned to Cs^+ in the silicate sheets.

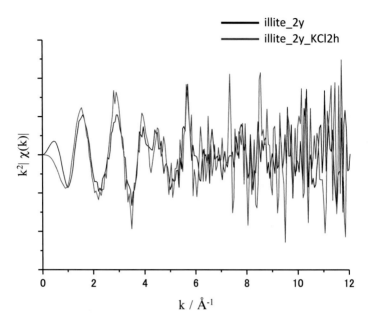

Fig. 1.2 k^2-weighted K-edge EXAFS spectra for illite_2y and illite_2y_KCl2h

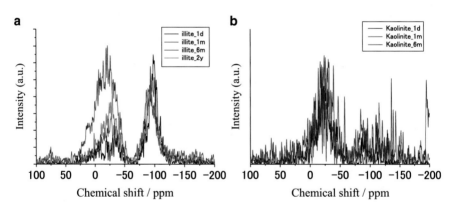

Fig. 1.3 NMR spectra of clays immersed in CsCl aqueous solution for several perids (**a**) illite, (**b**) kaolinite. The chemical shift reference is CsCl (aq)

These assignments agree well with the results of previous NMR experiments; Cs$^+$ with larger CN values exhibits a large field shift, while that with smaller CN values exhibits a small field shift [13]. In another NMR experiment, there also existed two kinds of peaks for illite immersed in CsCl [14]. These results also support our assignment.

After re-ion-exchange (using KCl) over a relatively long period (2 days) for illite (Fig. 1.4), no peak was observed in the NMR spectrum, thereby indicating that all

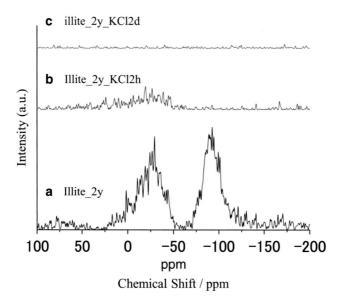

Fig. 1.4 NMR spectra of illite (**a**) immersion in CsCl solution for 2 years, (**b**) ion-exchanged illite by immersion in KCl for 2 h, (**c**) ion-exchanged illite by immersion in KCl for 2 days. The chemical shift reference is CsCl (aq)

the Cs^+ ions on the surface and in the silicate sheet had been replaced by K^+. This result supports our assignment because Cs^+ on the clay surface and in the silicate sheet is easily ion-exchanged by K^+ [6]. This result also indicates that no FESs were observed in this experiment. One of the reasons for the absence of the FES signature is the low concentration of the total amount of Cs^+. After re-ion-exchange for a short period of time (2 h), we observed only one peak at −30 ppm. This result can be attributed to the fact that Cs^+ in the silicate sheet is more easily replaced by K^+ than that on the surface.

In conclusion, solid-state ^{133}Cs NMR is useful for analyzing Cs^+ adsorbed onto clay minerals; this method can be used to distinguish Cs^+ sites in the clay minerals. In the near future, we plan to perform structure analysis of Cs^+ in soil in order to understand the mechanism of transfer of Cs from the soil to plants. We believe that this study will contribute to the recovery of agriculture in Fukushima in the near future.

1.6 Conclusions

We performed ^{133}Cs NMR experiments in conjunction with XRD and EXAFS to analyze the adsorption of Cs^+ onto clay minerals such as illite and kaolinite. Our NMR results indicate that the observed peaks at −30 and −100 ppm can be assigned to Cs^+ on the surface and in the silicate sheet, respectively. After re-ion-exchange by using aqueous KCl, the Cs^+ ions were replaced by K^+ ions. We believe that our findings will contribute to a better understanding of the mechanism of Cs transfers from the soil to plants. We plan to use the NMR method in our future studies on understanding of the mechanism of transfer of Cs^+ from the soil to plants.

Acknowledgments This work was supported by the Collaborative Research Program of Institute for Chemical Research, Kyoto University (No. 2015- 68, 70), Research Institute for Sustainable Humanosphere, Kyoto University, and Kureha Trading Co., Ltd. The synchrotron radiation experiments were performed at the BL14B2 beamline of the SPring-8 facility with the approval of the Japan Synchrotron Radiation Research Institute (JASRI) (Proposal No. 2015A1662). We would like to thank Kenji Hara, Hiroshi Goto and Hironori OFuchi for fruitful discussions.

References

1. IAEA (2006) Environmental consequences of the Chernobyl accident and their remediation: twenty years of experience. International Atomic Energy Agency, Vienna
2. Fujimura S, Muramatsu Y, Ohno T, Saitou M, Suzuki Y, Kobayashi T, Yoshioka K, Ueda Y (2015) Accumulation of (137)Cs by rice grown in four types of soil contaminated by the Fukushima Dai-ichi Nuclear Power Plant accident in 2011 and 2012. J Environ Radioact 140:59–64
3. Kato N, Kihou N, Fujimura S, Ikeba M, Miyazaki N, Saito Y, Eguchi T, Itoh S (2015) Potassium fertilizer and other materials as countermeasures to reduce radiocesium levels in rice: results of urgent experiments in 2011 responding to the Fukushima Daiichi Nuclear Power Plant accident. Soil Sci Plant Nutr 61:179–190
4. Ehlken S, Kirchner G (2002) Environmental processes affecting plant root uptake of radioactive trace elements and variability of transfer factor data: a review. J Environ Radioact 58:97–112
5. Tsukada H, Nakamura Y (1999) Transfer of ^{137}Cs and stable Cs from soil to potato in agricultural fields. Sci Total Environ 228:111–120
6. Yamaguchi N, Takata Y, Hayashi K, Ishikawa S, Kuramata M, Eguchi S, Yoshikawa S, Sakaguchi A, Asada K, Wagai R, Makino T, Akahane I, Hiradate S (2012) Behavior of radiocaesium in soil-plant systems and its controlling factor (in Japanese). Bull Nat Inst Agro-Environ Sci 31:75–129
7. Fuller AJ, Shaw S, Ward MB, Haigh SJ, Mosselmans JFW, Peacock CL, Stackhouse S, Dent AJ, Trivedi D, Burke IT (2015) Caesium incorporation and retention in illite interlayers. Appl Clay Sci 108:128–134
8. Okumura M, Nakamura H, Machida M (2014) First-principles studies of cesium adsorption to frayed edge sites of micaceous clay minerals. In: Abstract paper of the 248thAmericanchemicalsocietynationalmeeting&exposition, San Francisco
9. Sato K, Fujimoto K, Dai W, Hunger M (2013) Molecular mechanism of heavily adhesive Cs: why radioactive Cs is not decontaminated from soil. J Phys Chem C 117:14075–14080
10. Cremers A, Elsen A, Depreter P, Maes A (1988) Quantitative-analysis of radiocesium retention in soils. Nature 335:247–249

11. Fan Q, Yamaguchi N, Tanaka M, Tsukada H, Takahashi Y (2014) Relationship between the adsorption species of cesium and radiocesium interception potential in soils and minerals: an EXAFS study. J Environ Radioact 138:92–100
12. Bostick BC, Vairavamurthy MA, Karthikeyan KG, Chorover J (2002) Cesium adsorption on clay minerals: an EXAFS spectroscopic investigation. Environ Sci Technol 36:2670–2676
13. Minami T, Tokuda Y, Masai H, Ueda Y, Ono Y, Fujimura S, Yoko T (2014) Structural analysis of alkali cations in mixed alkali silicate glasses by ^{23}Na and ^{133}Cs MAS NMR. J Asian Ceram Soc 2:333–338
14. Kim Y, Cygan RT, Kirkpatrick RJ (1996) ^{133}Cs NMR and XPS investigation of cesium adsorbed on clay minerals and related phases. Geochim Cosmochim Acta 60:1041–1052
15. Kim Y, Kirkpatrick RJ (1997) ^{23}Na and ^{133}Cs NMR study of cation adsorption on mineral surfaces: local environments, dynamics, and effects of mixed cations. Geochim Cosmochim Acta 61:5199–5208
16. Kim Y, Kirkpatrick RJ, Cygan RT (1996) ^{133}Cs NMR study of cesium on the surfaces of kaolinite and illite. Geochim Cosmochim Acta 60:4059–4074
17. Maekawa A, Momoshima N, Sugihara S, Ohzawa R, Nakama A (2014) Analysis of ^{134}Cs and ^{137}Cs distribution in soil of Fukushima prefecture and their specific adsorption on clay minerals. J Radioanal Nucl Chem 303:1485–1489
18. Ravel B, Newville M (2005) ATHENA, ARTEMIS, HEPHAESTUS: data analysis for X-ray absorption spectroscopy using IFEFFIT. J Synchrotron Radiat 12:537–541
19. Yaita T (2013) Interpretation of Cs adsorption behavior based on the EXAFS, TR-Dxafs, and STXM methods. Mineral Mag 77:2532
20. Yaita T, McKinley I (2013) Fundamental approaches toward development of radiocesium removal methods from soil and the other related materials, waste reduction and management optimization. http://fukushima.jaea.go.jp/initiatives/cat01/pdf00/07__Yaita.pdf

Chapter 2
Speciation of ^{137}Cs and ^{129}I in Soil After the Fukushima NPP Accident

Tomoko Ohta, Yasunori Mahara, Satoshi Fukutani, Takumi Kubota, Hiroyuki Matsuzaki, Yuji Shibahara, Toshifumi Igarashi, Ryoko Fujiyoshi, Naoko Watanabe, and Tamotsu Kozaki

Abstract We evaluated the migration of radionuclides (^{131}I, ^{129}I, ^{134}Cs, ^{136}Cs, ^{137}Cs, and ^{132}Te) in the surface soil after the Fukushima nuclear accident. The radionuclides in the soil collected late March in 2011 were barely leached with ultrapure water, indicating that these are insoluble. We observed the chemical behavior of ^{137}Cs and ^{129}I in soil: (1) ^{137}Cs was predominantly adsorbed within a depth of 2.5 cm from the ground surface; (2) ^{137}Cs was hardly released from soil by the water leaching experiments that lasted for 270 days; (3) approximately, more than 90 % of ^{137}Cs was adsorbed on organic matters and the residual fractions, while ^{129}I was mainly fixed on the Fe-Mn oxide and organically bounded fraction. Therefore, we conclude that ^{137}Cs and ^{129}I in soil seldom leach into the soil water and migrate downward because of the irreversible adsorption. The shallow groundwater which residence time is short.

Keywords Cesium-137 • Iodine-129 • Speciation • Soil • Fukushima nuclear accident • Migration • Groundwater

T. Ohta (✉) • T. Igarashi • R. Fujiyoshi • N. Watanabe • T. Kozaki
Faculty of Engineering, Hokkaido University, Nishi8, Kita13, Sapporo, Hokkaido 060-8628, Japan
e-mail: tomoohta@eng.hokudai.ac.jp

Y. Mahara
Kyoto University, Yosidahonmachi, Sakyou-ku, Kyoto-Shi, Kyoto 606-8501, Japan

The University Museum, The University of Tokyo, 16, 11, 2, Yayoi, Bunkyou-ku, Tokyo 113-0032, Japan

H. Matsuzaki
The University Museum, The University of Tokyo, 16, 11, 2, Yayoi, Bunkyou-ku, Tokyo 113-0032, Japan

S. Fukutani • T. Kubota • Y. Shibahara
Research Reactor Institute, Kyoto University, Kumatori, Sennangun, Osaka 590-0494, Japan

© The Author(s) 2016
T. Takahashi (ed.), *Radiological Issues for Fukushima's Revitalized Future*,
DOI 10.1007/978-4-431-55848-4_2

2.1   Introduction

A number of radionuclides (including ^{137}Cs, ^{134}Cs, ^{136}Cs, ^{131}I, ^{132}Te) were released into the atmosphere from the Fukushima Daiichi NPP accident in March, 2011. On March 15–17 and 21–23, deposition increased in the areas surrounding Fukushima prefecture because north–easterly, easterly, and south–easterly winds under a low-pressure system transported the radionuclides from the Fukushima NPP, and subsequent precipitation associated with the same system washed radioactive materials out of the radioactivity plume, thereby effectively depositing them on land [1–5]. In addition to the radioactive plume that covered the Fukushima Prefecture, two other large plumes suffered severe radioactive contamination over north Japan. Precipitation from these plumes caused high-radioactive spots across wide areas including the Tokyo metropolitan [1].

Although the eastern parts in the Tokyo metropolitan area are located far from the Fukushima NPP, high ^{137}Cs and ^{131}I deposition was observed in the areas. Therefore, the Tokyo metropolitan area is one of the hot spots of radioactive fallout from the Fukushima NPP accident [4]. The Tokyo metropolitan hot-spot area has a high-density population, and many residents have been worried about the radiation exposure from the Fukushima NPP. The released ^{131}I from the Fukushima NPP contaminated tap water with rainfall that precipitated in the Tokyo metropolitan hot-spot areas on March 23, 2011, whereas ^{131}I was also detected in the tap water of the Kanto and Tohoku regions through mid-April [1].

At present, ^{131}I released from the Fukushima NPP accident in March, 2011 is not detected in the environment, because its short half-life is only 8 days. On the contrary, ^{129}I with a half-life of 15.7 Ma is easily found. The ^{129}I/^{127}I ratio measured in surface soil could provide information on the local deposition of ^{131}I released from the Fukushima NPP, if the ratio of ^{131}I and ^{129}I before release from the broken reactors can be estimated [6, 7]. The ^{131}I behavior in the unsaturated zone and shallow groundwater immediately after the accident may be inferred from the ^{129}I content.

This study has two objectives: (1) to examine the behavior of ^{137}Cs and ^{131}I in the surface soil of the Kanto loam in the Tokyo metropolitan hot-spot areas and Fukushima immediately after the accident and (2) to determine the ^{137}Cs and ^{129}I speciation in the soil to evaluate ^{137}Cs and ^{131}I contamination in the shallow groundwater which residence time is short.

2.2   Material and Methods

2.2.1   Soil Samples

One surface-soil sample within a depth of 0.5 cm and four surface-soil samples within a depth of 1 cm were collected on March 29, 2011 at the western Tokyo metropolitan area (WTMA, W1) and on March, 30 and 31, 2011 at the eastern Tokyo metropolitan hot-spot area (ETMA, E1-4), respectively. The soil samples collected at western Tokyo were placed in a polypropylene vessel (vessel A), 5 cm in diameter and 10 cm in height, and each of the four soil samples collected at eastern Tokyo was packed into a polypropylene vessel (vessel B), 2.2 cm in diameter and 1.2 cm in height.

Surface-soil core samples (5 cm^2 × 10 cm) were collected at the WTMA on August 25, 2011 and ETMA, on October 14–18, 2011, respectively. The undisturbed soil cores were sampled using a cylindrical stainless steel core sampler, 0–10 cm in depth. The soil core samples were cut into the lengths of 0–2.5, 2.5–5 cm, and 5–10 or 0–2.5 and 2.5–5.0 cm. Each sample was stored in plastic vessels (vessels A). We also collected soil in the same manner in Nagadoro, Fukushima (N1) in May, 2012. Then we measured the vertical profiles of ^{137}Cs in soil.

2.2.2   Column-Infiltration Experiments Using the Rainwater from the Tokyo Metropolitan Hot-Spot Area

It rained in the southern Ibaraki Prefecture from 0:00 to 2:00 LT on March 31, 2011, resulting in a total precipitation of 5.5 mm in Tsukuba. A rainwater sample was collected during precipitation at the ETMA (Japan Meteorological Agency). The rainwater sample (101 mL) was placed in a polypropylene vessel (5 cm in diameter and 10 cm in height, vessel A).

We investigated the infiltration of ^{137}Cs from the rainwater in the soil environment via column experiments on April 1, 2011. Sand (Toyoura Standard Sand, Toyoura Keiseki Kogyo Co., Ltd., Yamaguchi, Japan) and soil were used in the columns. The soil, classified by the *World Reference Base* as a haplic stagnosol, was collected from a depth of approximately 30 cm from an outcrop at the campus of Kyoto University Research Reactor Institute (KURRI), Osaka, Japan. The soil contained a small amount of organic matter (organic carbon 2.7–2.8 %, pH 5.7–5.8) and was devoid of ^{137}Cs and ^{134}Cs.

The sand was rinsed several times to remove clay minerals using deep groundwater that did not contain ^{137}Cs. The sand and soil samples were packed into two columns (20 mm in diameter and approximately 1 cm in depth). After filling, columns A (sand) and B (soil) were soaked in pure water for 3 h. Next, the rainwater sample (50 mL) was passed through each column at a rate of 0.7 mL·min^{-1}, and then, each column was rinsed with ultrapure water (84 mL). Although the flow rate used was faster than the actual rate of rainwater infiltration in the Kanto district (approximately 1.5 m·y^{-1}), this provides a conservative estimation of radionuclide migration. After infiltration, the sand and soil samples were stored separately in polypropylene containers (vessel B). Then, these vessels were placed into a different container, and the radioactivity of the samples in each container was measured using a gamma-ray spectrometer.

Table 2.1 Concentrations of major ions in groundwater

Element	Concentration (mg·L^{-1})	Element	Concentration (mg·L^{-1})
Na	6.92	Cl$^-$	5.83
K	7.00	SO$_4{}^{2-}$	13.4
Ca	22.0	HCO$_3{}^-$	110
Mg	7.87	pH	8.3

2.2.3 Leaching of Radionuclides from Soils Using the Batch Method

Radioactivities of the 100-g soil samples collected on March 29, 2011 at WTMA were immediately measured by the following method. The soil samples were mixed with the ultrapure water and shaken by hand for 30 min. The mixture was stored for 15 min, and the solution was separated by ultra-centrifuge (Kokusan Co., Ltd, Japan), and then, radioactivity in the supernatant was measured by gamma-ray spectrometry.

Approximately 10 g of the surface-soil samples collected from the WETA were added to ultrapure and groundwater samples, and then, they were agitated at a speed of 100 rpm for 90 and 270 days. The groundwater sample was discharged from the Kanto-loam layer in the Tokyo metropolitan hot-spot area, and was sampled in March, 2010 before the Fukushima NPP accident. The concentrations of Na$^+$, K$^+$, Mg^{2+}, and Ca^{2+} in the groundwater were measured by cation chromatography and those of Cl$^-$, SO$_4{}^{2-}$ and NO$_3{}^-$ were measured by anion chromatography, whereas HCO$_3{}^-$ was measured by titration using HCl. Table 2.1 lists the major cation and anion of the groundwater samples.

After leaching, solid-liquid separation was conducted by centrifugation at 2000 rpm for 5 min. Then, the solution was filtered through a membrane filter with a pore size of 0.45 μm (Advantech, Co., Ltd., Japan). After storing each fraction in a U-8 vessel, radioactivities of ^{131}I, ^{134}Cs, ^{136}Cs, ^{137}Cs, and ^{132}Te were measured by gamma-ray spectrometry.

2.2.4 Separation of ^{137}Cs and ^{129}I in Soil Samples

We extracted ^{137}Cs and ^{129}I from three surface soil samples from the Kanto-loam layer in the ETMA and the Nagadoro in Fukushima. The soil of depths of 0–2.5 cm (E5, E6, N1) was well mixed and approximately 10 g soil sample was taken by cone and quartering method. Approximately 10 g of the samples were used in the sequential-extraction experiment [8, 9]. A ratio of solution to sample of 5 (v/w) was used for extraction in each step.

Fraction 1: After ultrapure water was added to the soil sample, the suspension was shaken for 24 h at room temperature, and then the suspension was stored overnight. After extraction, solution was separated from the soil residue by centrifugation at 2000 rpm for 5 min. The solution was filtered through a membrane filter with a pore size of 0.45 μm (Advantech, Co., Ltd., Japan). The fraction of the filtrate represents water-soluble species. The remaining solid on the membrane filter was combined with the residue for the next leaching step.

Fraction 2: 1 M of NaAc was added to the residue from Fraction 1. The suspension was shaken for 12 h at room temperature and stored overnight. The fraction of the filtrate represents exchangeable species.

Fraction 3: 1-M NaAc – HAc (pH 5) was added to the residue from Fraction 2, and the suspension was shaken for 12 h at room temperature. The fraction of the filtrate represents carbonate-bound species.

Fraction 4: 0.04-M NH$_2$OH·HCl in 25 % (v/v) HAc (pH 2) was added to the residue from Fraction 3 and stirred in a hot-water bath at 80 °C for 4 h. The fraction of the filtrate represents species associated with solids via chemical-sorption mechanisms that can be released into the extraction solution with a weak reducing agent, and they mainly include species bound to Fe/Mn oxides.

Fraction 5: 30 % H$_2$O$_2$ was added to the residue, in which HNO$_3$ had already been added to adjust the final pH to 2. The suspension was agitated for 2 h at 85 °C. After the suspension was cooled to room temperature, 1.8-M NH$_4$Ac in 11 % HNO$_3$ (v/v) was added, and the extraction continued for 30 min at room temperature. The fraction of the filtrate is associated with organic matter.

After each fraction was stored into a U-8 vessel, ^{137}Cs was measured by gamma-ray spectrometry.

2.2.5 Purification of Iodine Isotopes for Accelerator Mass Spectrometry (AMS) Measurement

One-milliliter solution of nitric acid (Kanto Chemical Co., Ltd.) and 0.5 mL of H$_2$O$_2$ (Kanto Chemical Co., Ltd.) were added to sample solutions separated from each fraction. The dissolved iodine was oxidized to I$_2$ and was then separated from the sample solution into 10 mL of chloroform (Wako Co., Ltd.). The chloroform was separated from the sample solution, and then 10 mL of 0.1 M of NaHSO$_3$ (Wako Co., Ltd.) was added to the chloroform to extract I$^-$ into the NaHSO$_3$ solution. The NaHSO$_3$ solution was separated from the chloroform, and 1 mL of 6 M NaCl (Aldrich Co., Ltd.) was added to the solution. A 0.1 mL portion of 1 M AgNO$_3$ (Aldrich Co., Ltd.) was then added to the solution, and the solution was agitated, causing AgI to precipitate with AgCl. This precipitation was allowed to continue for 30 min before the mixture was centrifuged for 5 min at 3000 rpm. The mixture of AgCl and AgI precipitate was separated from the solution. AgCl was separated from AgI by adding 4 mL of concentrated NH$_3$ to dissolve AgCl only. The AgI precipitate was rinsed with 5 mL of ultrapure water to yield a pure AgI sample, which was then dried in an electric oven at 70 °C for 40 min. The AgI sample was added to Nb powder at an Nb and AgI ratio of 4 (w/w). The detail chemical separation was described by a previous paper [10]. ^{129}I and ^{127}I were measured by AMS (Malt, Tokyo Univ., Japan), and detailed procedure of measurement of ^{129}I/^{127}I atomic ratios were described by Matsuzaki et al. [11].

2.2.6 *Measurement of Radioactivity in Environmental Samples by Gamma-Ray Spectrometry*

The radioactivity of the rainwater and soil sample was measured using a p-type high-purity germanium detector (IGC-309, Princeton Gamma-Tech) with 40 % relative efficiency and a multichannel analyzer (7600-000, Seiko EG&G) with a high-voltage circuit (7600-310) and pulse height analyzer (7600-510) in KURRI. The gamma-ray counting efficiency of the detector was estimated by constructing a relative gamma-ray-counting-efficiency curve using a certified mixed-radionuclide gamma-ray reference source containing ^{57}Co (122.1 keV), ^{137}Cs (661.7 keV), and ^{60}Co (1173 and 1332 keV), which were normalized to the 1460-keV gamma-ray peak of ^{40}K in KCl. A quadratic function was fitted to the logarithmic relationship between the relative counting efficiency and gamma-ray energy using the least-squares method and was normalized at 1460 keV. The radiation energies of ^{131}I, ^{134}Cs, ^{136}Cs, ^{137}Cs, and ^{132}Te were 364, 796, 818, 662, and 228 keV, respectively. The sum effect of gamma rays from ^{134}Cs was corrected by measuring the ^{134}Cs solution. ^{134}Cs was produced by the neutron activation of CsCl at KURRI. CsCl was dissolved in the ultrapure water and the prepared ^{134}Cs solution. The detection limits of ^{131}I, ^{134}Cs, and ^{137}Cs with a measuring time of 10,000 s were 0.11, 0.099, and 0.12 Bq for vessel B and 0.20, 0.19, and 0.23 Bq for vessel A, respectively.

The radioactivity of undisturbed core samples from the ETMA and radio-Cs samples leaching into the ultrapure water and groundwater samples was measured in the Isotopes Centre, Hokkaido University using a p-type, high-purity germanium detector (model IGC-309, Princeton Gamma-Tech) with 40 % relative efficiency. The detection limits of ^{134}Cs and ^{137}Cs with a measuring time of 864,000 s were 0.02 and 0.02 Bq for the U-8 vessel, respectively.

2.3 Results and Discussion

Table 2.2 shows the concentrations of ^{134}Cs, ^{136}Cs, ^{137}Cs, ^{131}I, and ^{132}Te in the surface soils collected in March, 2011 from the Tokyo metropolitan hot-spot area (but only W6 was collected in August, 2011). Short half-lives for ^{136}Cs, ^{131}I, and ^{132}Te were detected in the surface-soil samples except W6. The radioactivities of

Table 2.2 Concentrations of ^{134}Cs, ^{136}Cs, ^{137}Cs, ^{131}I, and ^{132}Te in surface soils in Tokyo metropolitan area

Sampling site number	Sampling sites	Depth cm	Sampling date	^{134}Cs kBq·kg^{-1}	^{137}Cs kBq·kg^{-1}
W1	WTMA	0–0.5	March 29, 2011	0.517 ± 0.01	0.530 ± 0.07
E1	ETMA	0–1	March 30, 2011	2.86 ± 0.07	3.25 ± 0.07
E2	ETMA	0–1	March 30, 2011	2.70 ± 0.14	2.90 ± 0.12
E3	ETMA	0–1	March 31, 2011	1.01 ± 0.06	0.720 ± 0.06
E4	ETMA	0–1	March 31, 2011	1.80 ± 0.07	1.70 ± 0.18
W6	WTMA	0–5	Aug. 25, 2011	0.078 ± 0.002	0.093 ± 0.002
		^{131}I kBq·kg^{-1}	^{136}Cs kBq·kg^{-1}	^{132}Te kBq·kg^{-1}	^{134}Cs/^{137}Cs
W1	WTMA	3.43 ± 0.01[a]	0.051 ± 0.002[a]	0.373 ± 0.05[a]	1.03 ± 0.13
E1	ETMA	9.36 ± 0.10[b]	0.301 ± 0.360[b]	0.338 ± 0.050[b]	1.14 ± 0.03
E2	ETMA	11.1 ± 0.21[b]	0.220 ± 0.056[b]	0.168 ± 0.074[b]	1.07 ± 0.07
E3	ETMA	13.4 ± 0.2[b]	0.128 ± 0.050[b]	0.271 ± 0.056[b]	0.71 ± 0.10
E4	ETMA	7.32 ± 0.12[b]	0.632 ± 0.076[b]	0.328 ± 0.049[b]	0.94 ± 0.11
W6	WTMA		–	–	1.19 ± 0.03

WTMA Western Tokyo metropolitan area, *ETMA* Eastern Tokyo metropolitan area
[a]Correction at 5:00 March 29, 2011 at local time
[b]Correction at 2:00 March 31, 2011 at local time

^{131}I, ^{134}Cs, and ^{137}Cs in the surface-soil samples except W6 collected at ETMA were measured, and W6 measured only the radioactivities of ^{134}Cs and ^{137}Cs. ^{131}I was the source of the maximum radioactivity in the soil in March, 2011 in the Tokyo metropolitan hot-spot area. The concentration of ^{131}I ranged from 9.4 to 13 k Bq·kg^{-1} before rain (March 30, 2011) and from 7.2 to 11 k Bq·kg^{-1} after rain, whereas that of ^{137}Cs ranged from 0.72 to 3.3 k Bq·kg^{-1} before rain and 1.7–2.9 k Bq·kg^{-1} after rain (March 31). As the fallout radionuclides on the surface soils were mainly washed out from the atmospheric aerosol plume by precipitation, ^{137}Cs concentrations being similar to ^{131}I in the soil before rain on March 30 and after rain on March 31 suggest that the rainfall on March 21 and 22 (and the very small precipitation on March 15 and 16) was major source to most radionuclides in the soil.

The concentrations of ^{131}I, ^{134}Cs, and ^{137}Cs in the rainwater collected at the ETMA toward the end of March, 2011 were 66 ± 3, 28 ± 2, and 31 ± 2 Bq·L^{-1}, respectively. The concentration of each radionuclide was corrected at 2:00 LT, 31 March 2011.

Table 2.3 Radioactivities collected in sand and soil (column A and B), and collection efficiencies

Sample ID	Sand column	Soil column
Material	Sand	Soil
Weight (g)	3.2	4.7
^{131}I (Bq)	ND	0.82 ± 0.06
^{137}Cs (Bq)	1.44 ± 0.09	1.26 ± 0.08
^{131}I collection efficiency (%)	–	31 ± 3
^{137}Cs collection efficiency (%)	93 ± 6	81 ± 5

ND: below detection limit

Table 2.3 lists radioactivities trapped in both columns of sand and soil, and the ^{131}I- and ^{137}Cs-trapping efficiencies in both columns. In column A (sand), almost all of the ^{137}Cs was trapped on sand with the trapping efficiency of 93 ± 6 %. In column B (soil), the ^{137}Cs-trapping efficiency was 81 ± 5 %. Therefore, we considered almost all ^{137}Cs in rainwater to be adsorbed on sand and the Haplic Gray Upland soil within a depth of 1 cm from the ground, while the trapping efficiency of ^{131}I in both sand and soil columns was below the detection limit and 30 %, respectively. These data indicate that retardation of ^{131}I in both sand and soil downward was smaller than that of ^{137}Cs.

Table 2.4 shows the concentrations of ^{137}Cs and ^{134}Cs in the surface soil collected at the ETMA in October, 2011, 7 months after the accident. Almost all ^{134}Cs and ^{137}Cs in the Kanto-loam soil were within a depth of 2.5 cm from the ground. The column test and the distribution of ^{137}Cs obtained from the core sample in the field indicated that ^{134}Cs and ^{137}Cs were strongly adsorbed on the surface soil within a depth of several centimetres from the surface. Vertical profiles of radionuclides in soil in Koriyama, Fukushima, showed that more than 90 % of ^{131}I was found to be within about 5 cm depth from the surface in soil layer after the accident [12].

Following the leaching tests from the surface soil collected at WTMA into the pure water immediately after sampling in March, 2011, the leachate was separated from soil. ^{134}Cs, ^{137}Cs, ^{131}I, ^{136}Cs, and ^{132}Te in soil did not leach into the water (Table 2.5). Table 2.5 lists the amount of radioactive Cs in the leachate when the surface soil of the ETMA or WTMA was mixed with the ultrapure water and groundwater during 90 and 270 days. Both ^{137}Cs and ^{134}Cs in each soil of the Kanto-loam layer and the Nagadoro were not released by leaching with both the ultrapure water and groundwater after mixing during 90 and 270 days.

Table 2.6 shows the leaching fractions of ^{137}Cs from each soil in the Kanto-loam layer of the ETMA and in the Nagadoro, Fukushima according to the sequential-extraction procedure described previously. The amount of ^{137}Cs leached into water (F1) from the Kanto-loam soil and Fukushima soil was below the detection limit. This has been demonstrated by the leaching results of radioactive Cs from the soil by the ultrapure water and groundwater, as shown in Table 2.5. The ratio of exchangeable ^{137}Cs from the two soil samples (E5 and E6) was less than 1 %, with 50–60 % of ^{137}Cs remaining in the residue. Approximately 98 % of ^{137}Cs was in the fraction of Fe–Mn oxide, organic matter, and the residue. The ^{137}Cs in the water fraction (F1) is undetectable and this is the same results of the extraction

Table 2.4 Concentrations of ^{137}Cs and ^{134}Cs in surface soil collected in ETMA and Nagadoro

Sample No.	Depth cm	^{134}Cs kBq·m^{-2a}	^{137}Cs kBq·m^{-2}	^{134}Cs/^{137}Cs Activity ratio
E5	0.0–2.5	173 ± 4	175 ± 3	0.99 ± 0.03
	2.5–5.0	0.911 ± 0.059	0.722 ± 0.041	1.26 ± 0.11
	5.0–10	3.94 ± 0.15	3.66 ± 0.12	1.08 ± 0.05
Total	0.0–10	178 ± 1	179 ± 1	0.99 ± 0.01
E6	0.0–2.5	147 ± 4	133 ± 4	1.11 ± 0.04
	2.5–5.0	22.7 ± 0.8	20.3 ± 0.6	1.12 ± 0.05
	5.0–10	12.0 ± 1.0	16 ± 1	0.75 ± 0.08
Total	0.0–10	182 ± 1	169 ± 1	1.07 ± 0.01
	cm	MBq·m^{-2}	MBq·m^{-2}	Activity ratio
N1	0.0–1.0	26.6 ± 0.5	26.4 ± 0.3	1.01 ± 0.02
	1.0–2.0	17.3 ± 0.4	17.1 ± 0.2	1.01 ± 0.03
	2.0–3.0	18.6 ± 0.4	18.9 ± 0.2	0.98 ± 0.02
	3.0–4.0	5.50 ± 0.16	5.73 ± 0.10	0.98 ± 0.03
	4.0–5.0	2.76 ± 0.06	2.82 ± 0.04	0.98 ± 0.03
	5.0–10	1.44 ± 0.05	1.48 ± 0.03	0.97 ± 0.04
	10–15	0.08 ± 0.01	0.09 ± 0.01	0.90 ± 0.13
	15–20	ND	ND	–
	20–25	ND	ND	–
	25–29	ND	ND	–
Total	0.0–29	72.3 ± 0.7	72.5 ± 0.4	1.00 ± 0.01

[a]Correction on March 11, 2011

test obtained in Table 2.5. The small fraction of exchangeable ^{137}Cs suggests that immediately after the ^{137}Cs in rainwater dropping on the surface soil, ^{137}Cs and ^{134}Cs were strongly adsorbed onto the soil of the Kanto loam, and they were not readily leached into the soil water. The amount of ^{129}I in fraction of F1 and F2 were 0.7–9.5 % and 1–3.8 %, respectively; therefore, ^{131}I in unsaturated soil moved downward faster than ^{137}Cs. More than 90 % of ^{137}Cs was in the fraction of organic matter and the residue, while ^{129}I was mainly fixed by Fe-Mn oxidation and organically bound.

The estimated ^{137}Cs migration rate for the Nishiyama loam soil, which was obtained in situ at Nishiyama (Nagasaki), is 1.0 mm·y^{-1} [13]. This is considerably less than the rainwater infiltration rate of 2.5 m·y^{-1}. Furthermore, ^{137}Cs was not detected in the groundwater of the Nishiyama area, suggesting that ^{137}Cs has not yet migrated to the groundwater table [14]. We collected 4 L of shallow groundwater in the Kashiwa city in the ETMA on August 17, 2012. The groundwater was filtered through membrane filter with a pore size of 80, 3, and 0.45 μm. The groundwater after the filtration was gradually reduced to 5 mL by a mantle heater. Both ^{137}Cs and ^{134}Cs were not detected in the shallow groundwater and the suspended matters collected from the filter (Table 2.7).

Table 2.5 Extraction of cesium isotopes in soil by pure water and groundwater (in Bq)

Nuclides/media	Pure water	Pure water	Groundwater	Groundwater
Sampling site	W1	W6	W6	E5
^{134}Cs	ND (0.17)	ND (0.02)	ND (0.02)	ND (0.02)
^{137}Cs	ND (0.16)	ND (0.02)	ND (0.02)	ND (0.02)
^{131}I	ND (0.19)	–	–	–
^{136}Cs	ND (0.15)	–	–	–
^{132}Te	ND (0.14)	–	–	–
Period of extraction	30 min	90 day	90 day	90 day
Sampling date of soil	March 29, 2011	Aug. 25, 2011	Aug. 25, 2011	Aug. 25, 2011
Measuring time	10,000 s	864,000 s	864,000 s	864,000 s
Nuclides/media	Pure water	Pure water	Pure water	Groundwater
Sampling site	E5	E6	N1	N1
^{134}Cs	ND (0.02)	ND (0.02)	ND (0.02)	ND (0.02)
^{137}Cs	ND (0.02)	ND (0.02)	ND (0.02)	ND (0.02)
Period of extraction	270 day	270 day	90 day	270 day
Sampling date of soil	August 25, 2011	August 25, 2011	May 24, 2012	May 24, 2012
Measuring time	864,000 s	864,000 s	864,000 s	864,000 s

ND: below the detection limit, (): detection limit value

Table 2.6 Distribution of ^{137}Cs and ^{129}I in the soil dissolved in each fraction solution

Fraction	E5 (%)	E6 (%)	N1 (%)
^{137}Cs			
F1: Water	0.0	0.0	0.0
F2: Exchangeable	0.4	0.9	1.67
F3: Bound to carbonates	1.3	0.9	1.43
F4: Bound to Fe-Mn oxides	6.0	2.1	1.12
F5: Bound to organic matter	37.8	33.1	5.27
Residual	54.4	63.1	90.5
Total	100	100	100
^{129}I			
F1: Water	9.5	0.8	0.74
F2: Exchangeable	3.8	1.0	2.40
F3: Bound to carbonates	5.4	1.3	3.92
F4: Bound to Fe-Mn oxides	72	20.9	82.3
F5: Bound to organic matter	9.2	66.4	0.91
Residual	–	9.6	9.74
Total	100	100	100

Although the ^{131}I moving velocity deduced from that of ^{129}I was greater than that of ^{137}Cs, the ^{129}I migration rate is lower than the water infiltration rate, due to the ^{129}I absorption on soil in the unsaturated zone.

In Kanto loam, soil water reaches depths of 20–30 cm, considering a 1–1.5 m·y^{-1} infiltration rate, after 80 days corresponding to the time length of 10 times of the ^{131}I

Table 2.7 Concentrations of ^{137}Cs and ^{134}Cs in groundwater and radioactivities of suspended matter (SS) of the groundwater sampled at the Tokyo metropolitan area after the accident

Nuclides	^{131}I (Bq·kg^{-1})	^{137}Cs (Bq·kg^{-1})	^{134}Cs (Bq·kg^{-1})
Groundwater	ND (0.007)	ND (0.008)	ND (0.008)
SS (~80 μm)	ND	ND	ND
SS (80 μm ~ 3 μm)	ND	ND	ND
SS (3 μm ~ 0.45 μm)	ND	ND	ND

ND: below detection limit, (): detection limit value

half-lives. Mainly ^{137}Cs was detected in litters in forest [5] but ^{137}Cs was detected to depth of 10 cm in soil without litter [15]. Because rainy force is buffered in litters when the surface of soils has litters, rain is hard to directly enter to the deep part of the soil. Water-soluble ^{131}I would merely move downward to depths of 30–40 cm, even if ^{131}I penetrated a depth of 10 cm in the soil without litters and/or grass immediately after the accident. Therefore, ^{131}I could never reach the depths of 50 cm in the groundwater table.

2.4 Conclusion

The sequentially chemical fractionations of ^{129}I and ^{137}Cs in soil indicate that most part of ^{137}Cs and ^{129}I were insoluble. Traces of ^{131}I in the soil water did not reach the 50-cm depth by late June, 2011, corresponding to the time length of 10 times of ^{131}I half-lives after the Fukushima NPP accident. Therefore, shallow groundwater could be safely useful water resource after the accident.

Acknowledgment This study is partially supported by JST Initiatives for Atomic Energy Basic and Generic Strategic Research. We are deeply appreciative of Dr. Iimoto for scientific support and advice for the groundwater sampling.

References

1. Amano H, Akiyama M, Chunlei B, Kishimoto T, Kuroda T, Muroi T, Odaira T, Ohta Y, Takeda K, Watanabe Y, Morimoto T (2012) J Environ Radioact 111:42–52
2. Hirose K (2012) J Environ Radioact 111:13–17
3. Ohta T, Mahara Y, Kubota T, Fukutani S, Fujiwara K, Takamiya K, Yosinaga H, Mizuochi H, Igarashi T (2012) J Environ Radioact 111:38–41
4. Ohta T, Mahara Y, Kubota T, Igarashi T (2013) Ana Sci 29:941–947

5. Mahara Y, Ohta T, Ogawa H, Kumata A (2014) Sci Rep 4:Article number 7121. doi:10.1038/srep07121
6. Muramatsu Y, Matsuzaki H, Toyama C, Ohno T (2015) J Environ Radioact 139:344–350
7. Miyake Y, Matsuazaki H, Fujiwara H, Saito T, Yamagata H, Honda M, Muramatsu Y (2012) Geochem J 46:327
8. Oughton DH, Salbu B, Riise G, Lien HN, Østby GA, Nøren A (1992) Analyst 117:481–486
9. Riise G, Bjørnstad HE, Lien HN, Oughton DH, Salbu B (1990) J Radioanal Nucl Chem 142:531–538
10. Ohta T, Mahara Y, Kubota T, Abe T, Matsueda H, Tokunaga T, Sekimot S, Takamiya K, Fukutani S, Matsuzaki H (2013) Nucl Instrum Methods Phys Res B 294:559–562
11. Matsuzaki H, Muramatsu Y, Kato K, Yasumoto M, Nakano C (2007) Nucl Instrum Methods Phys Res B 259:721–726
12. Ohno T, Muramatsu Y, Miura Y, Oda K, Inagawa N, Ogawa H, Yamazaki A, Toyama C, Saito M (2012) Geochem J 46:287–295
13. Mahara Y (1993) J Environ Qual 22:722–730
14. Mahara Y, Miyahara S (1984) J Geophy Res 89:7931–7936
15. Multidisciplinary investigation on radiocesium fate and transport for safety assessment for interim storage and disposal of heterogeneous waste. The initiatives for atomic energy basic and generic strategic Research, JST, 2013, 240407 (in Japaneses). Initiatives for Atomic Energy Basic and Generic Strategic Research by the Ministry of Education, Culture, Sports, Science and Technology of Japan

Chapter 3
Isotopic Ratio of ^{135}Cs/^{137}Cs in Fukushima Environmental Samples Collected in 2011

Takumi Kubota, Yuji Shibahara, Tomoko Ohta, Satoshi Fukutani, Toshiyuki Fujii, Koichi Takamiya, Satoshi Mizuno, and Hajimu Yamana

Abstract The isotopic ratios of radioactive cesium derived from the Fukushima accident were determined by γ-spectrometry and thermal ionization mass spectrometry (TIMS). In order to ascertain the initial ratios at the time of the accident, environmental samples collected during 2011 were used for the analysis. Soil, litter, and seaweed were incinerated, and the cesium contained therein was adsorbed into ammonium phosphomolybdate powder. The cesium in the seawater was adsorbed into AMP-PAN resin (Eichrom Technologies, LLC); its recovery ratio was almost one without the carrier being added. Incinerated samples and the AMP-PAN resin were then measured by γ-spectrometry. The cesium solution recovered from adsorbers was subjected to TIMS measurements. The isotopic ratios of ^{134}Cs/^{137}Cs and ^{135}Cs/^{137}Cs were found to be independent of the type of sample in question, as well as the sampling location; the ratios were 0.07 and 0.36 (decay correction: 11 March 2011), respectively, which differ from the results of atmospheric nuclear tests (i.e., 0 and 2.7, respectively). This difference in the ratio of ^{135}Cs/^{137}Cs will contribute to estimations of the origin of radioactive contamination in the future.

Keywords Thermal ionization mass spectrometry • Fukushima accident • Environmental samples • Isotopic ratio of ^{135}Cs/^{137}Cs and ^{134}Cs/^{137}Cs

3.1 Introduction

The Fukushima Daiichi Nuclear Power Plant (FDNPP) disaster gave rise to the release of a large amount of radioactive material into the environment [1–3]. Radioactive cesium nuclides were spread and stored in east Japan because of

T. Kubota (✉) • Y. Shibahara • S. Fukutani • T. Fujii • K. Takamiya • H. Yamana
Research Reactor Institute, Kyoto University, Kumatori, Osaka 590-0494, Japan
e-mail: t_kubota@rri.kyoto-u.ac.jp

T. Ohta
Hokkaido University, Sapporo, Hokkaido 060-8628, Japan

S. Mizuno
Nuclear Power Safety Division, Fukushima Prefectural Government, Fukushima 960-8043, Japan

© The Author(s) 2016
T. Takahashi (ed.), *Radiological Issues for Fukushima's Revitalized Future*,
DOI 10.1007/978-4-431-55848-4_3

their high volatility and long half-life. However, global fallout from atmospheric nuclear tests was the main contaminant source before the Fukushima accident [4]. Yellow sand, which flies from China in the spring, is known to contain radioactivity from atmospheric nuclear tests [5]. In addition to this periodic input, it is possible that radioactive contaminants from the Fukushima accident may be present. Even though the radioactive materials are found somewhere in the future, the environment can be considered to be the same situation before the Fukushima accident as long as they originate from the global fallout; in other words, the environment is not contaminated by the Fukushima accident. The migration of contaminants from the Fukushima accident is an important factor in resolving the Fukushima accident. It is natural that radioactive materials will be found even in locations far from Fukushima, such as west Japan, because of the global fallout. It is therefore important, from the perspective of future studies on Fukushima, to determine whether such contaminants originate from the Fukushima accident or not.

The use of nuclear energy, both in terms of nuclear reactors and nuclear weapons, provides several radioactive cesium nuclides in its products. The different production history in the two processes, however, results in different isotopic ratios of cesium. ^{134}Cs is produced as a result of neutron irradiation of ^{133}Cs, which can originate as a fission product. On the other hand, the production amount of ^{134}Cs in nuclear tests can be ignored because nuclear reactions and neutron irradiation cease after an extremely short period of time. Accordingly, ^{134}Cs is considered to be contaminant in nuclear reactors. Although ^{135}Cs and ^{137}Cs are produced as direct fission products, the production of ^{135}Cs is affected by neutron irradiation conditions. Because ^{135}Xe, which is the parent nuclide of ^{135}Cs, has a large neutron capture cross section, the production of ^{135}Cs is suppressed in long periods of neutron irradiation; this results in the isotopic ratio of ^{135}Cs/^{137}Cs from nuclear reactors being different from that of nuclear tests. In this way, the isotopic ratios of Cs (e.g., ^{134}Cs/^{137}Cs and ^{135}Cs/^{137}Cs) can reveal the origin of contaminants in the environment.

Isotopic ratios can be determined in two ways: γ-spectrometry and mass spectrometry. It is easy to measure ^{134}Cs and ^{137}Cs by γ-spectrometry, a nondestructive analysis method. However, the ^{135}Cs of pure beta emitter cannot be measured, and the decay of ^{134}Cs (with a relatively short half-life) prohibits the isotopic analysis. On the other hand, mass spectrometry can determine the presence of ^{134}Cs, ^{135}Cs, and ^{137}Cs nuclides after chemical and/or physical purification. The measurement of ^{135}Cs permits isotopic analysis even after the ^{134}Cs has decayed out. Mass spectrometry can potentially be hindered by isobar nuclides; however, thermal ionization mass spectrometry (TIMS) with a high signal-to-noise ratio sufficiently suppresses the isobar effect. In the ionization process, isobar nuclides (in this study, cesium and barium) are separated according to their different ionization energies. We investigated various environmental samples by using TIMS [6], which is suitable for determining the isotopic ratios of radioactive cesium in the environment for long periods of time.

The isotopic ratio of ^{135}Cs/^{137}Cs determined by TIMS can estimate the origin of contamination in environmental samples; this process requires the ratio at the time of Fukushima accident to be used as the initial value. The initial values in various samples – not only soil and litter but also seawater and seaweed – are important factors in discussing the migration and mixing of radioactive cesium released in the Fukushima accident, other accidental releases, and atmospheric nuclear tests. In this study, the isotopic measurement of the various samples collected from May to September 2011 was conducted by γ-spectrometry and TIMS.

3.2   Materials and Methods

The environmental samples were collected in Fukushima prefecture from May to September 2011, and four of them were subjected to analysis. A litter sample was collected along Route 399 in Iitate village, where severe contamination was observed. Seawater and seaweed samples were collected at the Matsushita beach in Iwaki city. A soil sample was collected in Hinoemata village, which is one of the Fukushima local government's sampling plots.

Before analyzing the specific activity and isotopic ratios of cesium, the samples collected were treated as follows. The soil sample was dried at 105 °C and sieved through a 2 mm screen in order to remove pebbles, tree roots, and leaves; it was then incinerated at 450 °C in order to disintegrate organic matter. The litter sample was dried, cleaned by hand, and then incinerated. The seaweed sample was dried and incinerated without cleaning. An aliquot of these incinerated samples was dissolved in HNO_3 and then purified with ammonium phosphomolybdate (AMP) [6] in order to recover the cesium fraction. The seawater sample was treated with AMP-PAN in order to concentrate the cesium, which was eluted with NH_4OH.

The radioactivity of ^{134}Cs and ^{137}Cs in incinerated samples and AMP-PAN was measured by γ-spectrometry. The radioactivity was measured with a HPGe semiconductor detector, which was calibrated with a standard radioactive source of ^{137}Cs (662 keV) and ^{60}Co (1132 and 1337 keV) [7]. Each sample was placed apart from the detector, thereby reducing the coincidence summing effect in the determination of ^{134}Cs. In this case, the lower detection limit would rise; however, the radioactivity in samples was sufficiently high and could therefore be determined in this study. The isotopic ratios of ^{134}Cs/^{137}Cs and ^{135}Cs/^{137}Cs in the purified cesium fraction were measured by TIMS. The isotopic ratios of ^{134}Cs/^{137}Cs and ^{135}Cs/^{137}Cs were determined by TIMS, and a TaO activator was used to enhance the counting efficiency. The effect of isobaric barium can be ignored owing to the fact that it has a different ionization energy than cesium does [6].

3.3   Results and Discussion

In the analysis of radioactive cesium nuclides in seawater by γ-spectrometry, the natural cesium carrier is ordinarily added into the recovery process of cesium because of its low concentration. The addition of the carrier in mass spectrometry

Table 3.1 Specific activity of ^{137}Cs in AMP-PAN resin and the ^{137}Cs recovery ratio

	Weight (g)	Radioactivity of ^{137}Cs (Bq)	Specific activity (Bq/kg)	Recovery ratio
1st column	0.87	16.4 ± 0.2		
2nd column	1.49	12.8 ± 0.2		
Total (15 kg seawater)		29.2 ± 0.3	1.95 ± 0.02	0.97 ± 0.10
Initial [7]			2.0 ± 0.2	

Table 3.2 Radioactivity of ^{134}Cs and ^{137}Cs as determined by γ-spectrometry

Sample	Dry weight (g)	Radioactivity		Radioactivity ratio of ^{134}Cs/^{137}Cs	Collection date	Location
		^{134}Cs (Bq)	^{137}Cs (Bq)			
Soil	12.98	3.4 ± 0.3	4.0 ± 0.1	0.87 ± 0.07	15 September 2011	Hinoemata
Litter	15.90	6562 ± 57	6194 ± 43	1.06 ± 0.01	23 May 2011	Iitate
Seaweed	17.78	10.4 ± 0.5	10.7 ± 0.2	0.97 ± 0.05	25 August 2011	Iwaki
Seawater	AMP-PAN	28.9 ± 0.9	29.2 ± 0.3	0.99 ± 0.03	25 August 2011	Iwaki

could prevent to detection of the signal isotopes of interest by isotope dilution, thereby decreasing isotopic ratios below detection limit and covering with large signal from the carrier added. A seawater sample of known concentration was used in order to confirm qualitative recovery without adding the carrier. The seawater sample was treated with HNO_3 to pH 1 and then passed through two columns, connected in series and filled with AMP-PAN, at a rate of 5 kg/d. The specific activity of ^{137}Cs and the recovery ratio onto the AMP-PAN resin, listed in Table 3.1, show that an AMP-PAN resin of about 2.5 g is sufficient for completely extracting cesium isotopes from 15 kg of seawater.

The specific activity in environmental samples (as determined by γ-spectrometry) and the activity ratio of ^{134}Cs/^{137}Cs are listed in Table 3.2. The specific activity depends on the type of sample in question and the distance of the collection point from the FDNPP. However, the activity ratio shows that the contamination was derived from the Fukushima accident, although the activity ratio for soil collected at Hinoemata village is slightly low. The activity ratios were converted to isotopic ratios, which were then compared to the results of TIMS (see below).

The results of TIMS measurements are listed in Table 3.3, where the isotopic ratios evaluated from the results of γ-spectrometry are also listed. Among the four samples – litter, soil, seaweed, and seawater – only the litter sample yielded isotopic ratios for both ^{134}Cs/^{137}Cs and ^{135}Cs/^{137}Cs. In particular, the intensity of the signal from the soil sample was too low to evaluate both isotopic ratios. Other elements obviously interfered with the mass spectrometry measurements of the seaweed sample, and thus neither isotopic ratio was evaluated. Both evaluation failures were probably caused by inadequate processes and/or the shortcomings of the purification

Table 3.3 Isotopic ratio of ^{134}Cs/^{137}Cs and ^{135}Cs/^{137}Cs as determined by γ-spectrometry and thermal ionization mass spectrometry (TIMS)

Sample	Isotopic ratio of ^{134}Cs/^{137}Cs		Isotopic ratio of ^{135}Cs/^{137}Cs	Distance from FDNPP	Reference
	γ-spectrometry	TIMS			
Soil	0.059 ± 0.005	Not detected	Not detected	WSW, 160 km	This work
Litter	0.073 ± 0.001	0.0691 ± 0.0014	0.3574 ± 0.0020	NW, 34 km	
Seaweed	0.067 ± 0.003	Not evaluated	Not evaluated	SSW, 55 km	
Seawater	0.068 ± 0.003	Not evaluated	0.3617 ± 0.0051	SSW, 55 km	
Grass	0.069 ± 0.009	0.0722 ± 0.0004	0.3622 ± 0.0006	NW, 42 km	[6]
Bark	0.067 ± 0.002	0.0698 ± 0.0005	0.3663 ± 0.0005	NNW, 3 km	
Root	0.069 ± 0.001	0.0684 ± 0.0008	0.3586 ± 0.0008	WNW, 8 km	
Moss	0.070 ± 0.000	0.0713 ± 0.0003	0.3663 ± 0.0007	SSW, 2 km	

process, which first requires the purification of cesium. The mass spectrometry of the seawater sample shows that the ^{134}Cs signal overlapped with the tail of the ^{133}Cs signal (see Fig. 3.1); nevertheless, the isotopic ratio of ^{135}Cs/^{137}Cs can be obtained. Despite the fact that it is based on only one result of the litter sample, the isotopic ratio of ^{134}Cs/^{137}Cs evaluated from TIMS is in good agreement with that from γ-spectrometry, thereby demonstrating the validity of both measurement methods. The isotopic ratios of ^{135}Cs/^{137}Cs of the litter and the seawater sample collected in 2011 were both 0.36 (decay correction: 11 March 2011).

The results obtained in this study are now compared to those of other references. Those of the previous work [6] are listed in Table 3.3. The isotopic ratio of ^{134}Cs/^{137}Cs in the samples (evaluated from both γ-spectrometry and TIMS) agrees with the results of the other study; however, the ratio of soil from Hinoemata is slightly lower. Furthermore, the isotopic ratios of ^{135}Cs/^{137}Cs in land samples are in good agreement with that in the seawater sample, which did not yield an isotopic ratio for ^{134}Cs/^{137}Cs. In another study [8], the isotopic ratios of ^{135}Cs/^{137}Cs in litter, lichen, and soil samples were determined with inductively coupled plasma tandem mass spectrometry (ICP-MS/MS) to be 0.333–0.375, depending on location and distance from the FDNPP. These three studies showed that the environmental samples (both land and marine samples) collected from the area remarkably contaminated by the Fukushima accident had isotopic ratios of ^{134}Cs/^{137}Cs of 0.07 and ^{135}Cs/^{137}Cs of 0.36, which are quite different from the values that arise from global fallout; such values were mainly the result of atmospheric nuclear tests and were almost 0 and about 2.7, respectively [6]. The radioactivity of ^{134}Cs will decrease to 1/30 of its initial value 10 y after the Fukushima accident, which will lead to difficulties in detecting ^{134}Cs. However, ^{135}Cs and ^{137}Cs can be sufficiently detected by TIMS at that time, and thus the isotopic ratio of ^{135}Cs/^{137}Cs related to the Fukushima accident and global fallout will become 0.44 and 3.4, respectively. The value of the isotopic ratio in environmental samples collected in the future can therefore demonstrate the origin of radioactive cesium by TIMS.

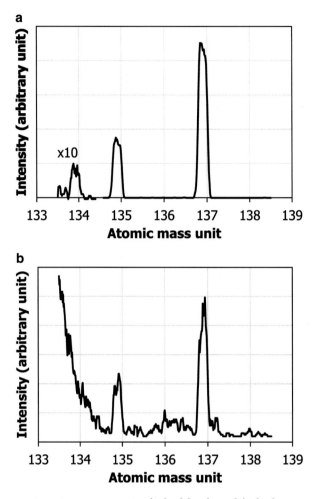

Fig. 3.1 Mass spectrometry measurements obtained by thermal ionization mass spectrometry (TIMS) for (**a**) litter and (**b**) sea water

3.4   Conclusion

In order to obtain the initial isotopic ratios of cesium at the time of the Fukushima accident, environmental samples were collected in Fukushima prefecture during 2011. The samples were treated with AMP and AMP-PAN in order to purify and recover the cesium; using the latter technique, the cesium contained in seawater was almost completely recovered. The isotopic ratios of cesium were determined by γ-spectrometry and TIMS. These results were in good agreement with each other (i.e., the results were independent of the determination method used) and with other studies as well. However, further development in the purification of the seaweed sample is required. On the basis of the initial isotopic ratio derived from the Fukushima accident, the isotopic ratio of cesium in environmental samples collected in the future can be expected to estimate the contribution of the accident through comparisons with the effects of global fallout.

Acknowledgements We wish to thank Mr. Mitsuyuki Konno and Mr. Satoru Matsuzaki for their help in collecting environmental samples and the Matsushimaya Ryokan staff for their support. This work was supported by the KUR Research Program for Scientific Basis of Nuclear Safety.

References

1. Hirose K (2012) 2011 Fukushima Dai-ichi nuclear power plant accident: summary of regional radioactive deposition monitoring results. J Environ Radioact 111:13–17
2. Ohta T et al (2012) Prediction of groundwater contamination with ^{137}Cs and ^{131}I from the Fukushima nuclear accident in the Kanto district. J Environ Radioact 111:38–41
3. Tagami K et al (2011) Specific activity and activity ratios of radionuclides in soil collected about 20 km from the Fukushima Daiichi Nuclear Power Plant: radionuclide release to the south and southwest. Sci Total Environ 409:4885–4888
4. Hirose K et al (2008) Analysis of the 50-year records of the atmospheric deposition of long-lived radionuclides in Japan. App Radiat Isot 66:1675–1678
5. Fukuyama T et al (2008) Contribution of Asian dust to atmospheric deposition of radioactive cesium (^{137}Cs). Sci Total Environ 405:389–395
6. Shibahara Y et al (2014) Analysis of cesium isotope compositions in environmental samples by thermal ionization mass spectrometry – 1. A preliminary study for source analysis of radioactive contamination in Fukushima prefecture. J Nucl Sci Technol 51:575–579
7. Kubota T et al (2013) Removal of radioactive cesium, strontium, and iodine from natural waters using bentonite, zeolite, and activated carbon. J Radioanal Nucl Chem 296:981–984
8. Zheng J et al (2014) ^{135}Cs/^{137}Cs isotopic ratio as a new tracer of radiocesium released from the Fukushima nuclear accident. Environ Sci Tech 48:5433–5438

Chapter 4
Application of Mass Spectrometry for Analysis of Cesium and Strontium in Environmental Samples Obtained in Fukushima Prefecture

Analysis of Cesium Isotope Compositions in Environmental Samples by Thermal Ionization Mass Spectrometry-2

Yuji Shibahara, Takumi Kubota, Satoshi Fukutani, Toshiyuki Fujii, Koichi Takamiya, Tomoko Ohta, Tomoyuki Shibata, Masako Yoshikawa, Mitsuyuki Konno, Satoshi Mizuno, and Hajimu Yamana

Abstract For the assessment of Fukushima Daiichi Nuclear Power Plant accident, the applicability of the thermal ionization mass spectrometry (TIMS), which is a type of mass spectrometry, was studied. For the study of the recovery/analysis method of cesium and strontium, at first, the radioactive cesium and strontium were generated by the irradiation of natural uranium at KUR. After this study, the applicability of this method to the environmental samples obtained in Fukushima prefecture was verified.

Keywords Fukushima Dai-ichi Nuclear Power Plant accident • Strontium • Cesium • Chromatography • Mass spectrometry • Isotopic ratio

Y. Shibahara (✉) • T. Kubota • S. Fukutani • T. Fujii • K. Takamiya • H. Yamana
Kyoto University Research Reactor Institute, 2, Asashiro-Nishi, Kumatori-cho, Sennan, Osaka 590-0494, Japan
e-mail: y-shibahara@rri.kyoto-u.ac.jp

T. Ohta
Hokkaido University, Kita 8, Nishi 5, Kita-ku, Sapporo, Hokkaido 060-8628, Japan

T. Shibata • M. Yoshikawa
Kyoto University Institute for Geothermal Sciences, Noguchi-baru, Beppu, Oita 874-0903, Japan

M. Konno • S. Mizuno
Nuclear Power Safety Division, Fukushima Prefectural Government, 8-2 Nakamachi, Fukushima, Fukushima 960-8043, Japan

© The Author(s) 2016
T. Takahashi (ed.), *Radiological Issues for Fukushima's Revitalized Future*,
DOI 10.1007/978-4-431-55848-4_4

4.1 Introduction

On the accident of Fukushima Daiichi Nuclear Power Plant (FDNPP), fission products (FP) such as radioactive Cs and Sr were widely released. The amounts of FP generated in each reactor were calculated by using ORIGEN code [1]. Many studies of radioactive Cs and Sr were performed to estimate external and internal exposures and to analyze the source of radioactive nuclides. These studies were typically performed by γ-ray spectrometry of ^{134}Cs ($T_{1/2} = 2.06$ y) and ^{137}Cs ($T_{1/2} = 30.2$ y) for the analysis of radioactive Cs and by β-spectrometry of ^{90}Sr ($T_{1/2} = 28.9$ y) for that of radioactive Sr.

In addition to ^{134}Cs and ^{137}Cs, radioactive ^{135}Cs ($T_{1/2} = 2.3 \times 10^6$) is also generated during the operation of reactors. Because of the difference in the generation process and the half-life of radioactive Cs, the isotopic ratios of ^{134}Cs/^{137}Cs and ^{135}Cs/^{137}Cs have been used for analyzing the operations of nuclear facilities [2–6]. Naturally occurring Sr has four stable isotopes (^{84}Sr, ^{86}Sr, ^{87}Sr, and ^{88}Sr), on the other hand, and the isotopic composition of Sr generated in reactors [1] are totally different from the natural abundance [7]. From the analysis data of the isotopic compositions, thus, the information on the origin of radioactive nuclide release would be obtained. The mass spectrometry provides the isotopic compositions of elements. Although mass spectrometry has been used for the analysis of radioactive nuclides and actinides, few studies have reported the analysis of radioactive Cs and Sr.

The purpose of the present study is to analyze Cs and Sr isotopes in environmental samples in Fukushima prefecture for source analysis and safety assessment. Although the amounts of radioactive Cs and Sr released in this accident were very huge, the contaminated environmental samples show the small radioactivity per unit weight of the contaminated environmental samples, since the contaminated area is very wide. For the study of the recovery/analysis method of cesium and strontium, at first, the radioactive Cs and Sr were generated by the irradiation of natural uranium at KUR. After this study, the applicability of this method to the environmental samples obtained in Fukushima prefecture was verified.

4.2 Experimental

4.2.1 Irradiation of UO₂ for Study of Radioactive Cs and Sr

10 mg of UO$_2$ of natural uranium was irradiated for 3 h at the Kyoto University Research Reactor with the neutron flux 5.5×10^{12} n/s cm^2. From the calculation with ORIGEN-II code [8], the amounts of the major radionuclide of Cs and Sr were estimated as 7.4×10^{-11} g (^{137}Cs) and 4.5×10^{-11} g (^{90}Sr), respectively. After standing for $ca.$ 2 days, radioactive Cs and Sr were recovered and analyzed.

4.2.2 Recovery of Cs and Sr

4.2.2.1 Isolation of TRU Elements

Cs and Sr were recovered with UTEVA^{TM}-resin (100–150 μm, Eichrom Technologies), Sr-resin (100–150 μm, Eichrom Technologies), ammonium phosphomolybdate (AMP), the cation exchange resin DOWEX^{TM} 50WX8 (100–200 mesh), and the anion exchange resin DOWEX^{TM} 1 X 8 (100–200 mesh).

The irradiated UO_2 was dissolved in 8 M HNO_3 (TAMAPURE-AA-100) and was evaporated to dryness at 403 K. 8 M HNO_3 was added and the insoluble residues removed by centrifugation. After centrifugation, H_2O_2 (TAMAPURE-AA 100) was added for the preparation of 8 M HNO_3/0.3 % H_2O_2 sample solution to isolate TRU elements such as U and Pu by the extraction chromatography with UTEVA-resin [9].

Three milliliter of the UTEVA-resin conditioned with diluted nitric acid was filled into a column of 54 mm in length and 6.5 mm in diameter and pretreated with 10 mL of 8 M HNO_3/0.3 % H_2O_2 before loading the solution. After loading the solution, the UTEVA-resin was rinsed with 8 M HNO_3 to elute alkaline earth metal elements [10]. The effluent was evaporated to dryness and dissolved in 10 mL of 3 M HNO_3 solution for the extraction chromatography with Sr-resin.

4.2.2.2 Recovery of Strontium

The solution was loaded to the Sr-resin conditioned with diluted nitric acid and filled into a column of 54 mm in length and 6.5 mm in diameter up to 3 mL. This effluent was evaporated at 403 K and the residue dissolved in 0.05 M HNO_3 for the recovery of Cs. After washing of the Sr-resin with 3 M HNO_3, Sr was recovered with 20 mL of 0.05 M HNO_3, evaporated to dryness, and dissolved in 10 μL of 1 M HNO_3.

4.2.2.3 Recovery of Cesium

After adding of 0.1 g of AMP to the Cs solution and stirring for several hours, the supernatant was removed from the mixed solution by centrifugation. A 20 mL 3 M ammonium hydroxide (TAMAPURE-AA 100) solution was used to dissolve the residue for subsequent anion-exchange ion chromatography.

After the final conditioning [11], a 3 mL portion of the anion-exchange resin was added to a column of 54 mm in length and 6.5 mm in diameter. The sample solution was added to the column, and the resulting eluate was collected and heated to dryness. The residue was dissolved in 20 mL of 0.1 M HNO_3 for the final purification with the cation-exchange ion chromatography.

The cation-exchange resin conditioned with hydrofluoric acid (TAMAPURE-AA-100), etc. [12] was filled into a column of 42 mm in length and 5.0 mm in diameter up to 1.5 mL. After loading the sample solution, the resin was washed with diluted nitric acid followed by 20 mL of 1.5 M HCl (TAMAPURE-AA 100) to recover Cs. The effluent was heated to dryness, and the residue was dissolved in 20 μL of 1 M HNO_3 for the analysis of the isotopic composition of Cs.

4.2.3 Analysis of Isotopic Composition of Cesium and Strontium

Isotopic compositions of Cs and Sr were measured with a TIMS (Triton-T1, Thermo Fisher Scientific). A 1 μL aliquot of each solution was loaded onto a rhenium filament with a TaO activator [13]. The standard material of SRM987 [14] was used as a reference material of mass spectrometry of Sr. The mass spectra of radioactive Cs and Sr were obtained with a secondary electron multiplier detector (SEM) because of the low total amounts of radionuclide loaded onto the filament.

4.2.4 Analysis of Environmental Samples

The plant samples were obtained from the south area of Iitate village, the northeast area of Okuma town, the southeast area of Futaba town, and southwest area of Futaba town in Fukushima prefecture from November 2012 and May 2013 (Table 4.1). The samples were washed three times with pure water and dried at 373 K. About 2.5 g of the dried samples was incinerated with a ring furnace at 873 K and dissolved in concentrated HNO_3 at 403 K and evaporated to dryness. 20 mL of 8 HNO_3 was added and the insoluble residues removed by centrifugation for the preparation of recovery of Cs and Sr. Recovery of Cs and Sr from environmental samples was also carried out by the same manner described above.

The concentration of [88]Sr was measured with an inductively coupled quadrupole mass spectrometer (ICP-QMS, HP-4500, Yokoagawa) and radioactivity of ^{90}Sr by Cherenkov counting [15]. The total concentration of radioactive Cs was measured by γ-spectrometry. The sample solutions were prepared as 50 ppm of [88]Sr in 1 M HNO_3 for the analysis of Sr and 5000 Bq/mL for ^{137}Cs in 1 M HNO_3 for the analysis of Cs. The mass spectra of radioactive Cs and Sr were obtained with a SEM, while those of stable Cs and Sr were obtained with Faraday cup detector, since the amounts of stable nuclide were much larger than those of radionuclide.

Table 4.1 List of samples and results of $^{87}Sr/^{86}Sr$ isotopic ratio measurement

Sampling area	Sample ID	Type	$\delta_{87/86}$[a]	Remarks
Iitate village (37.61 N, 140.80E)	ITT01	Grass (*Artemisia indica*)	−3.28(01)	*ITT01 to 07 were prepared by division of one sample*
	ITT02		−3.04(04)	
	ITT03		−3.20(09)	
	ITT04		−3.05(07)	
	ITT05		−3.11(07)	
	ITT06		−3.13(08)	
	ITT07[b]		−3.14(04)	
		ITT-av	−3.14(06)	
Okuma town (37.41 N, 141.03E)	OKM01[b]	Moss	−1.42(12)	
	OKM02[c]	Moss	−1.83(05)	
	OKM03	Bark (*Metasequoia glyptostroboides*)	−4.42(08)	
Futaba town-1 (37.45 N, 141.62E)	FTB01[b]	Bark (*Cryptomeria japonica*)	−2.51(08)	
	FTB02	Leaves of tree (*Camellia japonica*)	−3.75(09)	
	FTB03	Leaves of tree	−3.87(15)	*Same tree (Cryptomeria japonica), 03: attached leaves; 04: fallen leaves*
	FTB04	Leaves of tree	−4.14(09)	
	FTB05	Grass (*Artemisia indica*)	−3.29(09)	
	FTB06		−4.23(08)	
Futaba town-2 (37.45 N, 140.94E)	FTB35R[b]	Grass(*Artemisia indica*)	−2.96(08)	*Same grass, 35R: roots; 35 L: leaves*
	FTB35L		−4.30(08)	
Austria	IAEA-156	Grass (*Clover*)	−2.27(03)	

[a]Parentheses means experimental error in ±2 s.d
[b]Isotopic ratio of radioactive Cs has been reported in our previous study [11]
[c]Isotopic ratio of radioactive Cs was analyzed in this study

4.3 Results and Discussion

4.3.1 Isotopic Analysis of Radioactive Cs and Sr from Irradiated UO₂

Figure 4.1a shows the mass spectra of Cs recovered from the irradiated UO_2. In this measurement, ^{135}Cs, ^{136}Cs, and ^{137}Cs were detected: ^{134}Cs was not detected, because of the difference in the generation scheme. The observed isotopic ratios of ^{135}Cs/^{137}Cs and ^{136}Cs/^{137}Cs were obtained as 0.9103 ± 0.0008 and 0.00022 ± 0.00001. From our calculation with ORIGEN-II code [8], the loading amounts of ^{135}Cs, ^{136}Cs, and ^{137}Cs in this time were about 3.5, 0.03, and 3.7 pg respectively. This means that the femtogram level of Cs is detectable by TIMS.

Figure 4.2 shows the mass spectra of Sr both of stable (a) and radioactive (b) isotopes. At the measurement of 2.6 days later, ^{89}Sr, ^{90}Sr, and ^{91}Sr were detected. From our calculation with ORIGEN-II code [8], the loading amounts of ^{89}Sr, ^{90}Sr, and ^{91}Sr in this time were about 3, 4, and 0.04 pg respectively. This means that the femtogram level of Sr is also detectable by TIMS.

The measured isotopic ratios were 0.80 for ^{89}Sr/^{90}Sr and 0.01 for ^{91}Sr/^{90}Sr showing the agreement with the calculated value (0.79 for ^{89}Sr/^{90}Sr and 0.01 for ^{91}Sr/^{90}Sr). Because the half-life of ^{91}Sr is 9.5 h, the mass spectrum of ^{91}Sr disappeared at the measurement of 31 days later. The measured isotopic ratio of ^{89}Sr/^{90}Sr is 0.53 showing the agreement with the calculated value of 0.54. At the measurement of 574 days later, only the mass spectrum of ^{90}Sr was observed because the half-life of ^{89}Sr is 50.5 days. This means that ^{89}Sr/^{90}Sr could not be analyzed by using a typical mass spectrometer after Sep. 2012, if we obtain the sample containing femtogram level of ^{90}Sr. The isotopic ratio of ^{90}Sr/^{stable}Sr would be therefore needed for our purpose.

Fig. 4.1 Mass spectra of Cs. (**a**) Recovered from UO₂ irradiated in KUR. (**b**) Recovered from environmental sample from Fukushima prefecture (Reproduced from Ref [11])

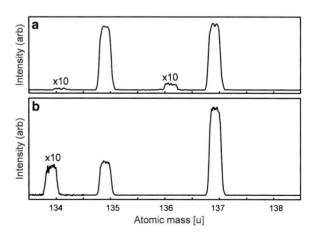

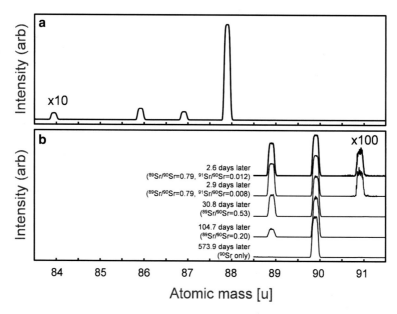

Fig. 4.2 Mass spectra of Sr. Stable isotopes (**a**) were obtained by measurement of Sr of SRM987 with a Faraday cup detector, and radioactive isotopes (**b**) were obtained by measurement of Sr recovered from UO_2 irradiated in KUR with a secondary electron multiplier detector

4.3.2 Analysis of Isotopic Compositions of Cs and Sr from Environmental Samples

4.3.2.1 Analysis of Cs

Figure 4.1b shows three peaks, representing ^{134}Cs, ^{135}Cs, and ^{137}Cs, were observed on the typical mass spectra of Cs recovered from environmental samples obtained in Fukushima prefecture [11], while the peak representing ^{136}Cs was not observed because of the half-life ($T_{1/2} = 13.2$ d). From the calculation with ORIGEN-II code [1], the isotopic ratio of ^{136}Cs/^{137}Cs in the fuel was estimated as *ca.* 0.00032. This value shows the same order compared with that of the irradiated UO_2, suggesting that we could obtain the three isotopic ratios of ^{134}Cs/^{137}Cs, ^{135}Cs/^{137}Cs, and ^{136}Cs/^{137}Cs until July 2011. Since there are three reactors in FDNPP, three isotopic ratios would bring the important information for the source analysis of radioactive Cs in the contaminated area in Fukushima prefecture.

Although we could not obtain the isotopic ratio of ^{136}Cs/^{137}Cs after July 2011, we can obtain the two-dimensional map with the isotopic ratios of ^{134}Cs/^{137}Cs and ^{135}Cs/^{137}Cs as shown in Fig. 4.3. All of the isotopic ratios of ^{135}Cs/^{137}Cs showed less than 0.4. This value was also much smaller than reported isotopic ratios of global fallout (~0.5 for Chernobyl accident and ~2.7 for nuclear weapon testing, corrected

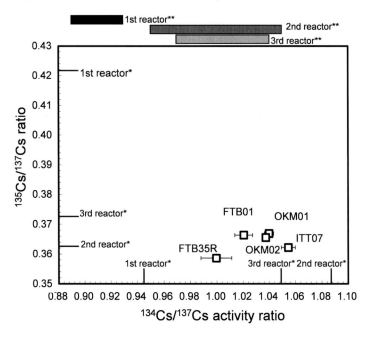

Fig. 4.3 ¹³⁵Cs/¹³⁷Cs (atomic ratio) vs ¹³⁴Cs/¹³⁷Cs (activity ratio). Error here means ±2SE. Data of OKM01, FTB01, FTB35R and ITT07 were reproduced from Ref [11]. Both of isotopic ratio was corrected to March 11, 2011. Single asterisk (*) represents calculation results from estimation of radioactive nuclides with ORIGEN-II code [1]. Double asterisk (**) represents values reported for ¹³⁴Cs/¹³⁷Cs activity ratio in polluted water [22]

to March 11, 2011 [11]) and the long half-life of ^{135}Cs ($T_{1/2} = 2.3 \times 10^6$ y), meaning that only the isotopic ratio of ^{135}Cs/^{137}Cs would also provide the information for the origin of radioactive Cs among Chernobyl accident, nuclear weapon testing, and FDNPP accident for the long term.

4.3.2.2 Analysis of Sr

The FP of Sr in each reactor has mainly five isotopes [1]: two stable isotopes of ^{86}Sr and ^{88}Sr and three radioactive isotopes of ^{89}Sr, ^{90}Sr, and ^{91}Sr. The relationship between the isotopic ratio of radioactive Cs and that of Sr estimated by ORIGEN Code calculation [1] is plotted in Fig. 4.4. In addition to the radioactive isotopes, the stable isotopes of Sr generated in each reactor show the characteristic profile. This suggests that the stable isotopes of Sr could be also used for the analysis of the FP of Sr.

Among the isotopic ratios of stable isotopes, the isotopic ratio of ^{87}Sr/^{86}Sr is important in the field of the geological chronology [16], because ^{87}Sr is generated by the β-decay of ^{87}Rb having the half-life of 4.9×10^{10} y. Thus, the isotopic ratio of stable isotopes, in this study, will be focused on the isotopic ratio of ^{87}Sr/^{86}Sr.

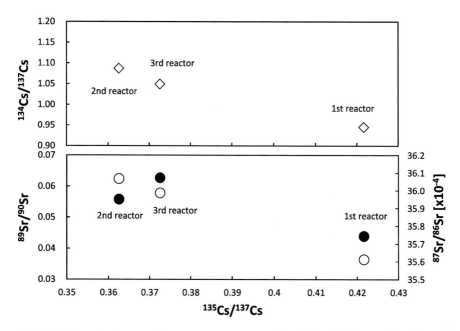

Fig. 4.4 Estimated relationship between isotopic ratios of ^{134}Cs/^{137}Cs, ^{89}Sr/^{90}Sr and ^{87}Sr/^{86}Sr and that of ^{135}Cs/^{137}Cs. Isotopic ratios were estimated by calculation results with ORIGEN Code [1]. *Open circle* means isotopic ratio of ^{89}Sr/^{90}Sr. *Closed circle* represents isotopic ratio of ^{87}Sr/^{86}Sr

The certified value for SRM987 of the isotopic ratio of ^{87}Sr/^{86}Sr showing the 95 % confidence intervals is 0.71036 ± 0.00026 [14]. The averaged measurement value was obtained as 0.71025 ± 0.00002 ($n = 26$) showing the agreement with the certified value.

In this study, the variations in the isotopic ratio of ^{87}Sr/^{86}Sr were normalized with that of SRM987; this would be expressed as delta-value ($\delta_{87/86}$) in per mill notation as the following equation:

$$\delta_{87/86} = \left(\frac{(^{87}Sr/^{86}Sr)\, sample}{(^{87}Sr/^{86}Sr)\, SRM987} \right) \times 1000.$$

The samples of ITT01 to ITT07 were prepared by the division of one sample. The $\delta_{87/86}$–values of samples ITT01 to ITT07 in Table 4.1 agreed within the error showing the reproducibility of the isotopic ratio measurement including chemical treatment. From the $\delta_{87/86}$–values of samples ITT01 to ITT07, the averaged $\delta_{87/86}$–value of them was obtained to be $\delta_{87/86} = -3.14 \pm 0.06\,‰$.

The results of the isotopic ratio measurements for all samples are summarized in Table 4.1 and shown in Fig. 4.5a. The result of the measurement for the reference material of IAEA-156: Radionuclides in clover [17] is also included. This reference material contains *ca.* 0.0075 Bq/g in June 2015. The $\delta_{87/86}$-values of the samples of

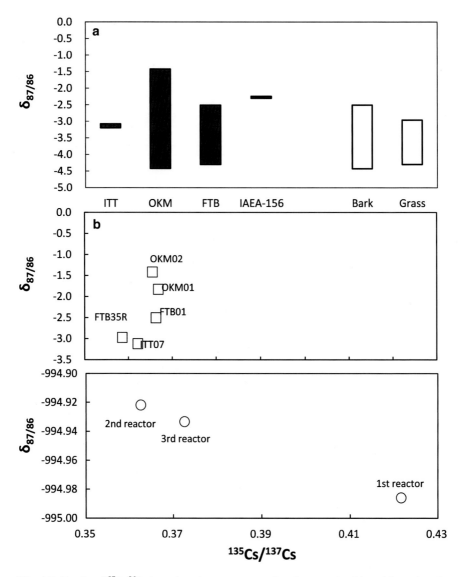

Fig. 4.5 Results of ^{87}Sr/^{86}Sr isotopic ratio measurement for plant samples (**a**), and isotopic ratio of ^{87}Sr/^{86}Sr as a function of ^{135}Cs/^{137}Cs (**b**). Bark is results of comparison between OKM03 and FTB01. Grass shows results of comparison between FTB35R and FTB35L. Isotopic ratio of ^{135}Cs/^{137}Cs of OKM01, FTB01, FTB35R and ITT07 were reproduced from Ref [11], and used after time correlation on March 11, 2011. *Open square* and *open circle* mean analytical results in this study and results of estimation by calculation results with ORIGEN Code [1]

Okuma range from -1.4 to -4.4, while those of Futaba range from -2.5 to -4.2. It is found that these values have significant difference, by comparison with the $\delta_{87/86}$-value of Iitate samples.

Though the samples OKM03 and FTB01 are bark samples from the plants of the same family, these showed different magnitudes (Fig. 4.5a and Table 4.1). The isotopic ratio of $^{87}Sr/^{86}Sr$ has received attention as the indicator of the production region of plants and reported the $\delta_{87/86}$–values ranged from -25.0 to 5.5 [18]. As the reason of the difference in the $\delta_{87/86}$–values among samples OKM03 and FTB01, two origins could be considered: the first is the difference in the $\delta_{87/86}$–values of soils of sampling point (as the supply source of Sr) and the second is the difference in the degree of the isotope fractionation during the translocation process (considered as the reason of the difference in the isotopic ratio between the parts of the identical organism). Because of the comparison of the $\delta_{87/86}$–values of the same parts in this case, the difference in the $\delta_{87/86}$–value among samples OKM03 and FTB01 might be caused by the soils in sampling area.

If the difference of $\delta_{87/86}$–values between samples OKM03 and FTB01 originated from a difference of contamination level by the FP of Sr, the isotopic ratio may show a correlation as

$$\left([^{87}Sr]_{OKM03}/[^{86}Sr]_{OKM03}\right) = \left([^{87}Sr]_{nat}/[^{86}Sr]_{nat}\right) \times (1-X)$$
$$+ \left([^{87}Sr]_{FP}/[^{86}Sr]_{FP}\right) \times X,$$

$$\left([^{87}Sr]_{FTB01}/[^{86}Sr]_{FTB01}\right) = \left([^{87}Sr]_{nat}/[^{86}Sr]_{nat}\right) \times (1-Y)$$
$$+ \left([^{87}Sr]_{FP}/[^{86}Sr]_{FP}\right) \times Y.$$

According to the relation and the concentrations of Sr; 72 ppm for OKM03 and 24 ppm for FTB03, the amount of the FP of ^{86}Sr contained in the sample OKM03 would be higher than that of FTB01, about 10.3 ng. This is equivalent to $ca.$ 10.5 μg of ^{90}Sr ($ca.$ 5.3×10^7 Bq) according to the averaged isotopic ratio of $^{90}Sr/^{86}Sr$ of the FP of Sr [1]. ^{90}Sr was not found in the plant samples by TIMS and Cherenkov counting having the detection limit of several ten mBq/g [15], however, suggesting that our samples contain $^{90}Sr <<10$ fg and was less than 1 Bq/g.

Sample FTB35R is roots, while FTB35L is leaves, of the same plant. The $\delta_{87/86}$-values (Fig. 4.5a and Table 4.1) showed a significant difference. Sample FTB35R shows higher $\delta_{87/86}$-value compared with sample FTB35L. The isotopic fractionations were observed in some biological processes. For example, the isotopic analysis of Sr [19], Fe [20], and Zn [21] proves that roots are isotopically heavy compared with the aerial parts; the maximum $\delta_{87/86}$-value was $ca.$ -5.0 for Sr, the maximum $\delta_{56/54}$-value was $ca.$ -1.4 for Fe, and the maximum $\delta_{66/64}$-value was $ca.$ -0.26 for Zn, respectively. Since the Cherenkov counting showed the amounts of ^{90}Sr in these samples were under the detection limit, the difference in the $\delta_{87/86}$-value between samples FTB35R and FTB35L might be caused by the isotopic fractionations in the biological processes along with the contamination of sample FTB35R by the soil.

The isotopic ratios of radioactive Cs in samples FTB01, OKM01, FTB35R, and ITT07 measured by TIMS have been reported in our previous study [11], and that in OKM02 was measured in this study. The relationships between the isotopic ratio of $^{87}Sr/^{86}Sr$ as $\delta_{87/86}$-value of these samples and that of $^{135}Cs/^{137}Cs$ are plotted in Fig. 4.5b. The isotopic ratios of $^{135}Cs/^{137}Cs$ show the significant difference from the reported values of the global fallout (*ca.* 0.5 for Chernobyl accident and *ca.* 2.7 for nuclear weapon testing corrected on March 11, 2011 [11]), while these values agreed with the estimated values with the results of ORIGEN Code calculation [1]. This means that all of the samples are contaminated by radioactive Cs released from FDNPP. The $\delta_{87/86}$-values of these samples, on the other hand, are far from that of the FP calculated by ORIGEN Code [1]. This suggests that the amount of deposit of 90Sr is very little compared with that of Cs and agrees with our previous report [15].

Although ^{90}Sr was not found in the plant samples suggesting that our samples contain $^{90}Sr \ll 10$ fg, typical mass spectrometers have the external analytical precision of ppm level. Assumed that this precision could be applied for the isotopic ratio of $^{90}Sr/^{stables}Sr$, the isotopic ratio of $^{90}Sr/^{stable}Sr$ must be higher than 10^{-6}. For the natural sample, since the Sr concentration ranges from ppb level to several hundred ppm level (Fig. 4.6), the detectable lower limit of the isotopic ratio of $^{90}Sr/^{stable}Sr$ can be evaluated.

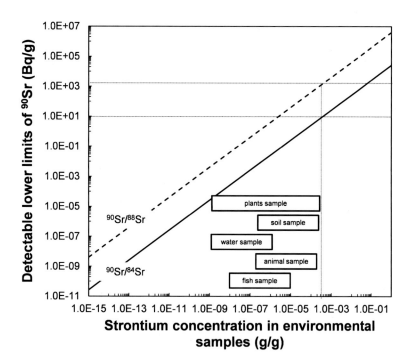

Fig. 4.6 Detectable lower limits of ^{90}Sr in environmental samples with TIMS. *Solid line* indicates a limit for $^{90}Sr/^{84}Sr$, and *broken line* a limit for $^{90}Sr/^{88}Sr$

If ^{88}Sr having the natural abundance *ca.* 82 % was used as reference isotope, the concentration of ^{90}Sr should be higher than 1 Bq/g in almost any type of sample. When the isotopic ratio of ^{90}Sr/^{84}Sr is used, because the abundance of ^{84}Sr (*ca.* 0.56 %) is lower than that of ^{88}Sr, the applicable range will become much wider than the case of ^{88}Sr (Fig. 4.5). The improvement in the sensitivity of ^{90}Sr detection and the obtaining of samples including small amounts of natural Sr will also bring wide applicable range.

4.4   Conclusions

Cs and Sr recovered from samples were analyzed by TIMS to study the applicability of TIMS for safety assessment and source analysis.

For the study of the recovery/analysis method of Cs and Sr, Cs and Sr were recovered from the natural uranium irradiated at KUR. From the measurement of radionuclide recovered from irradiated UO_2, it was concluded that several tens of femtogram level of radionuclide is detectable.

Cs and Sr were recovered from the environmental samples obtained from Fukushima prefecture and were analyzed by a method based on the results of irradiated UO_2. In the case of the analysis of Cs, it was confirmed that the analysis of the radioactive Cs by TIMS would provide important information for the source analysis. The isotopic ratio of ^{135}Cs/^{137}Cs was useful for the precise evaluation of the radioactive Cs from FDNPP apart from that of global fallout after the radioactivity of ^{134}Cs became below the detection limit of γ-ray measurement.

In the case of the analysis of Sr, on the other hand, the presence of ^{90}Sr was not detected in any samples, while the changes in the isotopic ratios of ^{87}Sr/^{86}Sr were observed. From the discussion for the amount of the FP of Sr, it was conjectured that the changes in the isotopic ratios of ^{87}Sr/^{86}Sr might be brought by some isotopic fractionation in the biological processes. The evaluation of the detectable lower limit of the isotopic ratio of ^{90}Sr/^{stable}Sr suggests that the isotopic ratio of ^{90}Sr/^{84}Sr is the most suitable index to judge a source of radioactive Sr released during the accident of FDNPP by TIMS.

References

1. Nishihara K, Iwamoto H, Suyama K (2012) JAEA-data/code 2012–018 [in Japanese]
2. Karam LR, Pibida L, McMahon CA (2002) Appl Rad Isot 56:369–374
3. Pibida L, MacMahon CA, Busharw BA (2004) Appl Rad Isot 60:567–570

4. Taylor VF, Evans RD, Cornett RJ (2008) J Environ Radact 99:109–118
5. Delmore JE, Snyder DC, Tranter T, Mann NR (2011) J Environ Radact 102:1008–1011
6. Snyder DC, Delmore JE, Tranter T, Mann NR, Abbott ML, Olson JE (2012) J Environ Radact 110:46–52
7. Berglund M, Wieser ME (2011) Pure Appl Chem 83:397–410
8. Ludwig SB, Renier JP (1989) Standard- and extended-Burnup PWR and BWR reactor models for the ORIGEN2 Computer Code. Oak Ridge National Laboratory, ORNL/TM-11018
9. Shibahara Y, Kubota T, Fujii T, Fukutani S, Ohta T, Takamiya K, Okumura R, Mizuno S, Yamana H (2015) J Radioanal Nucl Chem 303:1421–1424
10. Lee MH, Park JH, Oh SY, Ahn HJ, Lee CH, Song K, Lee MS (2011) Talanta 86:99–102
11. Shibahara Y, Kubota T, Fujii T, Fukutani S, Ohta T, Takamiya K, Okumura R, Mizuno S, Yamana H (2014) J Nucl Sci Technol 51:575–579
12. Yoshikawa M, Nakamura E (1993) J Min Petr Econ Geol 88:548–561
13. Birck JL (1986) Chem Geol 56:73–83
14. https://www-s.nist.gov/srmors/view_detail.cfm?srm=987. Accessed 20 Aug 2014
15. Kubota T, Shibahara Y, Fujii T, Fukutani S, Ohta T, Takamiya K, Okumura R, Mizuno S, Yamana H (2015) J Radioanal Nucl Chem 303:39–46
16. Faure G, Powell JL (1972) Strontium isotope geology. Springer, New York
17. Strachnov V, Valkovic V, Dekner R (1991) Report on the intercomparison run IAEA-156: radionuclides in clover. International Atomic Energy Agency, Austria, IAEA/AL/035
18. Almeida CM, Vasconcelos MTSD (2001) J Anal At Spectrom 16:607–611
19. de Souza GF, Reynolds BC, Kiczka M, Bourdon B (2010) Geochim Gosmochim Acta 74:2596–2614
20. Moynier F, Fujii T, Wang K, Foriel J (2013) Compt Rend Geosci 345:230–240
21. Weiss DJ, Mason TFD, Zhao FJ, Kirk GJD, Coles BJ, Horstwood MSA (2005) New Phytol 165:703–710
22. Komori M, Shozugawa K, Nogawa N, Matsuo M (2013) Bunseki Kagaku 62:475–483 [in Japanese]

Chapter 5
Migration of Radioactive Cesium to Water from Grass and Fallen Leaves

Hirokuni Yamanishi, Masayo Inagaki, Genichiro Wakabayashi, Sin-ya Hohara, Tetsuo Itoh, and Michio Furukawa

Abstract The TEPCO Fukushima Daiichi Nuclear Power Plant accident in March 2011 led to high amounts of emitted radioactive Cs being deposited on land by both rainwater and snowfall. In addition, a significant amount of Cs was deposited on the surface of leaves, and after the accident, both trees and grasses absorbed radioactive Cs through their roots. In order to assess the effect on water sources, it is therefore important to evaluate the amount of radioactive Cs migrating to the water from both grass and fallen leaves.

A number of samples of clover, dandelion, and mugwort were collected from the Yamakiya elementary school in Kawamata-machi, Date-gun, Fukushima-ken in May 2013 and May 2014. Fallen leaves were also sampled from the wood adjoining the school. Measurement of the Cs content in water was carried out by placing the sample in water for over 400 days at 10–30 °C. The radioactive Cs content was measured using the HPGe detector. In the case of grass, the amount of migration to water was saturated after about 120 days. The saturation levels of migration rate to water varied with kinds of grass in the range of 0.2–0.8. The migration rate for fallen leaves was not larger than 0.13. In addition, after leaching from grass or fallen leaves into water, the absorption of radioactive Cs to soil was observed, and therefore, migration would be limited to a small area.

Keywords Fallen leaves • Grass • Migration • Radioactive cesium • Waste decontamination • Water

H. Yamanishi (✉) • M. Inagaki • G. Wakabayashi • S.-y. Hohara • T. Itoh
Atomic Energy Research Institute, Kinki University, 3-4-1 Kowakae, Higashi-osaka, Osaka 577-8502, Japan
e-mail: yamanisi@kindai.ac.jp

M. Furukawa
Mayor of Kawamata-machi, Fukushima-ken, 30, Gohyakuda, Kawamata-machi, Date-gun, Fukushima-ken 960-1492, Japan

© The Author(s) 2016
T. Takahashi (ed.), *Radiological Issues for Fukushima's Revitalized Future*,
DOI 10.1007/978-4-431-55848-4_5

5.1 Introduction

A large amount of radioactive material was released by the nuclear power plant accident at the TEPCO Fukushima Daiichi site (1F site) in March 2011 [1]. The radioactive particles were transported by the prevailing wind and were deposited on the ground by both rain and snowfall. Due to these factors, the level of radiation in surrounding areas has been raised through the presence of radioactive material on the ground. As radiocesium is a source of both external exposure and internal exposure, its environmental behavior is of particular interest. Radiocesium adheres strongly to clays in the surface soil [2], and a very small amount is absorbed upon contact with water. Radiocesium was also deposited on the surface of leaves. In addition, following the deposition on the soil surface, trees and grasses absorbed the radiocesium in the soil through their roots.

Due to their long half-lives, Cs-134 (2 years) and Cs-137 (30 years) are responsible for the ongoing presence of radioactive materials outside the 1F site, even 4 years after the event. In the contaminated area outside the 1F site, decontamination work is now underway on residential land spaces and public facilities. During this process, the topsoil is removed and collected, and dried grass and fallen leaves are collected as contaminated waste. The waste for decontamination is stored in a flexible impermeable 1-m^3 bag and kept in a temporary depot for subsequent decontamination.

We are particularly interested in determining the quantity of radiocesium that migrates to water from grasses, fallen leaves, and soil. The environmental behavior of radiocesium was previously studied following the Chernobyl nuclear plant accident [3]. However, it is difficult to apply the result to Fukushima, as the environmental conditions such as soil type, precipitation amount, and temperature are very different at the two sites. The annual rainfall is approximately 1166 mm in Fukushima but only 621 mm in Pripyat, the town of Chernobyl. Due to the higher quantity of rainfall in Fukushima, it is possible that radiocesium is transferred through water. In addition, it should be considered that the chemical form of radiocesium that deposits in Fukushima is different to that deposited in Chernobyl. It is therefore necessary to use samples collected in Fukushima to study the environmental behavior of radiocesium in this area.

In order to assess the effect on water sources, it is therefore considered important to evaluate the amount of radiocesium migrating to the water from grass and fallen leaves. We herein report a process for the determination of the Cs content in grass and fallen leaves and in water following leaching. Samples collected in Fukushima will be used for the studies. We expect that it will be possible to use the obtained results for the prediction of radiocesium behavior in the environment.

5.2 Materials and Methods

5.2.1 Sample Collection

Environmental samples for this study were collected from the grounds of the Yamakiya elementary school, Kawamata-machi, Fukushima-ken, Japan. Fallen radiocesium gave a high topsoil Cs-137 content of 50 Bq/g in May 2011 [4]. Accuracy was expected to be improved using grass collected from the contamination site, rather than grass grown in a laboratory. In addition, at the contamination site, fallen leaves contained radiocesium on their surface due to deposition. Samples of clover, dandelion, pine needles, and mugwort were collected in May 2013 and May 2014. Fallen leaves were also collected in the forest that adjoins the school. In May 2013 and May 2014, the measured ambient dose rates at 1 m above ground level are 0.5–1.2 and 0.2–0.8 μSv/h, respectively.

Kawamata-machi is located approximately 40 km to the northwest of the 1F site. Following the accident, as the estimated annual dose was expected to exceed 20 mSv, the Yamakiya area in the southeast of Kawamata-machi was appointed a planned evacuation zone on April 22, 2011. The staff at the Atomic Energy Research Institute of Kinki University investigated the environmental radiation in cooperation with Kawamata-machi and began the collection of data, which contributed to understanding the radiation situation and aided in the discussions relating to the implementation of control measures [4, 5].

5.2.2 Sample Preparation

In order to prepare the plant samples for measurement, they were first preprocessed. First, the roots were removed from the grass samples, as they held soil that was contaminated with radiocesium. Other samples were washed with water to remove the soil particles on the surface of the samples. After washing, the samples were dried either in air or in an oven at 60 °C. The dried samples (15 g) were chopped finely (<1 cm length) and were placed into a U-8 container for measurement. The radiocesium concentration in each sample was measured using a hyperpure germanium semiconductor detector (HPGe detector).

5.2.3 Radiocesium Migration to Water

The measured samples (15 g) were divided into four equally sized portions and each portion placed into a water-permeable bag (95 × 70 mm), which was weaved with composite fiber composed of polyethylene, polypropylene, and polyester. The four bags containing the samples were soaked in water (500 mL) in a plastic container ($18 \times 13 \times 5$ cm), which was tightly closed, and allowed to stand for over 400 days. After this time, the resulting contaminated water was filtered, collected, and weighed. As the radiocesium concentrations in the sample were low, it was desirable to remove as much water as possible from the sample to precisely measure radioactivity. This was initially achieved by means of an evaporating dish and mantle heater. The dried matter was scraped away from the surface of the dish by washing with water (10 mL) and was moved to a U-8 container. The sample was then dried further by evaporation using an infrared oven. The radiocesium contents of samples were measured using the HPGe detector to give the radioactivity of the radiocesium that migrated to water from the plant samples.

Using the sample collected in May 2014, a similar experiment was conducted, which compares the migration rate to water using both deionized water and rainwater. The soaking period was 10–135 days, and the temperature was controlled at 25 °C.

5.2.4   Radiocesium Deposition in Soil

Radiocesium migrates to water from grass by means of soaking. However, water is often in contact with soil, and as the clay minerals present in soil strongly adsorb radiocesium, it is expected that the radiocesium in water can be deposited in soil. We therefore decided to investigate this by exposing soil to a solution containing radiocesium for 3 days, and the variations in radiocesium concentration in the solutions were measured. The solution required for this experiment was prepared using fallen leaves gathered at the Yamakiya Elementary School. Radiocesium was leached from the leaves by means of an ultrasound bath. The radiocesium concentration was then increased by reduction in the water content using an evaporating dish and a heating mantle. The radiocesium concentration in the solution was determined using an HPGe detector. A solution containing radiocesium (100 mL) was added to samples of dry field soil (10 or 30 g), and the resulting mixture was left to stand at 20–25 °C for either 14 h or 3 days. After this time, the aqueous solution was filtered to remove all soil particles and collected. The resulting residual radiocesium concentrations of all four solutions (10 g, 14 h), (30 g, 14 h), (10 g, 3 days), and (30 g, 3 days) were measured using an HPGe detector.

5.3   Results and Discussion

We first examined the sample collected in May 2013. The radioactivities of Cs-137 in the dried sample (15 g) before soaking were calculated to be $5.9 + 0.2$ Bq, 2.6 ± 0.2 Bq, 18.5 ± 0.4 Bq, 3.0 ± 0.1 Bq, 102 ± 2 Bq, and 369 ± 3 Bq, for clover, pine needles, mugwort, dandelion, fallen leaves (broad leaf), and fallen leaves (needle leaf), respectively. The amounts of Cs-137 eluted from each sample

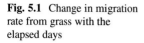

Fig. 5.1 Change in migration rate from grass with the elapsed days

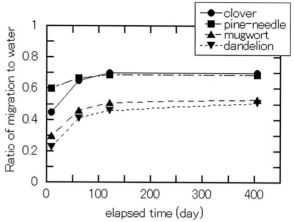

after soaking for 9 days were calculated to be 2.7 ± 0.1 Bq, 1.60 ± 0.04 Bq, 5.5 ± 0.2 Bq, 0.70 ± 0.03 Bq, 8.2 ± 0.2 Bq, and 6.3 ± 0.2 Bq, with an approximate 10-fold difference between the maximum and minimum values obtained. The migration rate is defined as the ratio of the radioactivity that migrated to water to the radioactivity of the sample before soaking. These migration rates were calculated to be 0.45 ± 0.03, 0.60 ± 0.04, 0.30 ± 0.01, 0.23 ± 0.01, 0.08 ± 0.002, and 0.017 ± 0.001. Fallen leaves have ratios in the range of 0.02–0.08 while grass exhibited larger ratios of 0.23–0.60.

After collection of the water, it was poured once more into the plastic container and was left to stand at 10–30 °C. This operation was repeated three times over different timeframes, more specifically, 50, 60, and 280 days, corresponding to 60, 120, and 400 days after beginning the soaking period, respectively. Figures 5.1 and 5.2 show the variation in migration rate over time. As shown in Fig. 5.1, in the case of grass, almost all the radiocesium which can be migrated had migrated to water after 60 days. In addition, it could be seen that after 120 days, little migration was observed, and so the total migration rate for 120 days was comparable to that for 400 days. In contrast, as shown in Fig. 5.2, the migration rate for fallen leaves was small, although an increase in migration rate even after 400 days was observed. This difference between grass and fallen leaves may be due to a number of reasons. Firstly, it must be considered that the radiocesium present in grass had been absorbed through the roots from the soil. This component of radiocesium might be present within the grass tissue due to weak adsorption, and thus may migrate to water. This would therefore result in a large migration rate for grass. In contrast, the radiocesium released following the accident was deposited on the surface of fallen leaves and was not absorbed through the roots. In addition, the radiocesium present on the fallen leaves did not migrate as easily to water, as the fallen leaves had been exposed to rain, and their surfaces washed before sampling. The migration rate shown in Fig. 5.2 is therefore small.

Fig. 5.2 Change in migration rate from fallen leaves and soil with the elapsed days

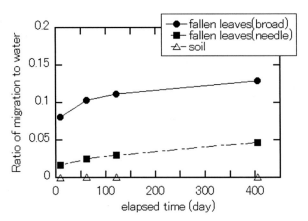

Figure 5.2 also shows the migration rate of radiocesium from dry soil (90 g) to water (500 mL). It was observed that the migration rate was very low. In addition, the migration rates increased over time, with a rate of $(3.5 \pm 0.1) \times 10^{-4}$ recorded after 10 days and $(1.2 \pm 0.1) \times 10^{-3}$ after 400 days.

Using the sample collected on May 2014, the migration rate for deionized water was compared with that obtained for rainwater. These samples were soaked in water for 10–135 days at 25 °C. As can be seen in Fig. 5.3, the differences between the two specifications of water were very small. Figures 5.4 and 5.5 show the variation in migration rate over time. The increasing trends of total migration have a similar pattern to Figs. 5.1 and 5.2. The migration rates from clover or pine needle were larger than that from mugwort, while the migration rate from fallen broad leaves was larger than that from fallen needle leaves.

Finally, Fig. 5.6 shows the results of the residual rate to solution (i.e., the ratio of concentration after soaking to concentration before soaking). The concentration of Cs-137 in the solution before the adsorption experiments was 0.177 ± 0.004 Bq/g. The residual rate in the case of 10 g soil was 0.29 ± 0.02 after soaking for 14 h and 0.17 ± 0.01 after soaking for 3 days. The residual rate in the case of 30 g soil was calculated as 0.11 ± 0.01 after soaking for 14 h and 0.07 ± 0.01 after soaking for 3 days. These data demonstrate that the residual rate decreases according to both the quantity of soil and according to the number of days in the soaking period. This is likely due to the large quantities of soil present in the area, and thus the radiocesium could not be transported through a large area. This would also not be possible even through migration to water, as the radiocesium would be rapidly adsorbed into the soil.

Fig. 5.3 Comparison of water, deionized water (dw) and rain water (rw)

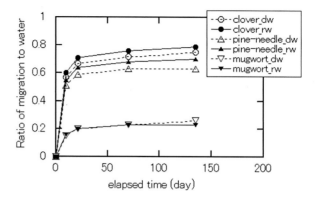

Fig. 5.4 Change in migration rate from grass with the elapsed days (using the sample May 2014)

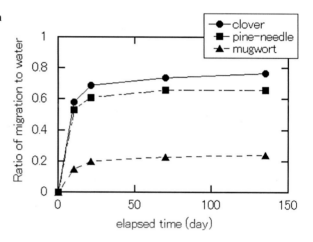

Fig. 5.5 Change in migration rate from fallen leaves and soil with the elapsed days (using the sample May 2014)

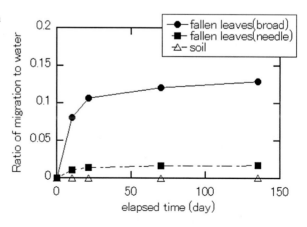

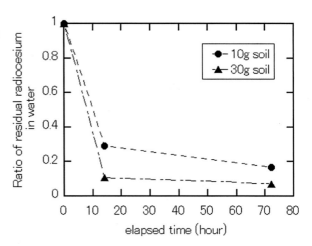

Fig. 5.6 Experimental result of the residual rate to solution

5.4 Conclusion

We obtained migration rates of radiocesium from samples of grass or fallen leaves (collected from Kawamata-machi, Fukushima-ken) soaked in water, with the rates for grass being larger than those for fallen leaves are. It is thought that the radiocesium present in grass exists in an easily transported form for rapid permeation within the grass tissue. In contrast, the radiocesium deposited on the surface of fallen leaves was exposed to rain for a long period before sampling, thus leading to a lower migration rate. As regarding grass and fallen leaves, there is no difference between at the test site and at other sites; the migration rates obtained in this study can be considered general values rather than specific values. We also investigated the effects of soaking contaminated soil in water and found that migration rates were low but increased gradually over time. In addition, it was observed that migration rates in deionized water were comparable with those in rainwater.

The adsorption of radiocesium into soil was studied using a solution containing radiocesium that had migrated to water from fallen leaves. The amount of radiocesium adsorbed by the soil depended on the quantity of soil. It is therefore expected that under natural conditions, the radiocesium, which is capable of migrating to water, can be easily adsorbed by the soil, and therefore cannot be transported over long distances. It may be possible to apply the results of this study for the prediction of radioactive cesium behavior in the environment.

References

1. Chino M, Nakayama H, Nagai H et al (2011) Preliminary estimation of release amount of I-131 and Cs-137 accidentally discharged from the Fukushima Daiichi Nuclear Power Plant into the atmosphere. J Nucl Sci Technol 48:1129–1134
2. Tsukada H, Toriyama K, Yamaguchi N et al (2011) Behavior of radionuclides in soil-plant system. Jpn J Soil Sci Plant Nutr 82:408–418 (in Japanese)
3. Report of the Chernobyl Forum Expert Group 'Environment' (2006) Environmental consequences of the chernobyl accident and their remediation: twenty years of experience. IAEA, Vienna
4. Yamanishi H, Hohara S, Wakabayashi G et al (2013) Survey of environmental radiation in Kawamata-machi, Fukushima-Prefecture. Radioisotopes 62:259–268, in Japanease
5. Itoh T, Furukawa M, Sugiura N et al (2011) Survey of environmental radiation in Kawamata-machi, Fukushima-ken. Annual report of Atomic Energy Research Institute, Kinki University 48:3–9 (in Japanese)

Chapter 6
Migration Behavior of Particulate ^{129}I in the Niida River System

Tetsuya Matsunaka, Kimikazu Sasa, Keisuke Sueki, Yuichi Onda,
Keisuke Taniguchi, Yoshifumi Wakiyama, Tsutomu Takahashi,
Masumi Matsumura, and Hiroyuki Matsuzaki

Abstract This study investigates the source and flux of particulate ^{129}I in the downstream reaches of the Niida River system in Fukushima. The upper watershed is a relatively highly contaminated zone located 30–40 km northwest of the Fukushima Daiichi Nuclear Power Plant. Samples of total suspended substance (SS) were collected continuously at Haramachi (5.5 km upstream from the river mouth) from December 2012 to January 2014 using a time-integrative SS sampler. Activity of ^{129}I and the ^{129}I/^{127}I ratio in SS were 0.9–4.1 mBq kg^{-1} and $(2.5$–$4.4) \times 10^{-8}$, respectively, and were strongly correlated with the total dry weight of SS samples with R^2 of 0.79–0.88. High SS ^{129}I activity and ^{129}I/^{127}I ratios were found in March, April, September, and October 2013. SS ^{129}I activity and ^{129}I/^{127}I ratios are considered to reflect the SS source, i.e., the more contaminated upper watershed or the less contaminated downstream area. The flux of particulate ^{129}I at the Haramachi site was estimated to be 7.6–9.0 kBq month^{-1} during September–October 2013. A relatively high amount of particulate ^{129}I may have been transported from the upstream to the downstream reaches of the Niida River by high rainfall over this period.

Keywords River system • Suspended substance • ^{129}I activity • ^{129}I/^{127}I ratio • Rain event • Migration behavior • Accelerator mass spectrometry

T. Matsunaka (✉) • K. Sasa • T. Takahashi • M. Matsumura
Tandem Accelerator Complex, University of Tsukuba, 1-1-1 Tennodai, Tsukuba,
Ibaraki 305-8577, Japan
e-mail: matsunaka@tac.tsukuba.ac.jp

K. Sueki • Y. Onda • K. Taniguchi • Y. Wakiyama
Center for Research in Isotope and Environmental Dynamics, University of Tsukuba,
1-1-1 Tennodai, Tsukuba, Ibaraki 305-8577, Japan

H. Matsuzaki
The University Museum, The University of Tokyo, Yayoi 2-11-16, Bunkyo-ku, Tokyo 113-0032,
Japan

© The Author(s) 2016
T. Takahashi (ed.), *Radiological Issues for Fukushima's Revitalized Future*,
DOI 10.1007/978-4-431-55848-4_6

6.1 Introduction

The nuclear accident at the Fukushima Daiichi Nuclear Power Plant (FDNPP), Japan, resulted in the massive release of high-volatility fission products to the environment, including ^{129}I ($T_{1/2} = 15.7$ million years) and ^{131}I ($T_{1/2} = 8.02$ days). The total amount of radionuclides discharged into the atmosphere was estimated to be 8.1 GBq for ^{129}I [1] and 120–150 PBq for ^{131}I [2, 3]. Approximately 13 % of the total amount of released ^{131}I was deposited over Japan via radioactive plumes [4]. Any short-lived ^{131}I deposited in the soil decays after a few months, however, long-lived ^{129}I derived from the FDNPP accident must be traced from land to the marine environment via river systems owing to its relatively high fission yield, high chemical reactivity, biological concentration in the marine ecosystem, and affinity for the thyroid gland although it is less radiologically harmful than ^{131}I.

This study aims to elucidate the source and flux of particulate ^{129}I in the downstream reaches of the Niida River, a small river in Fukushima Prefecture with an upper watershed located in a relatively high-contamination zone 30–40 km northwest of the FDNPP. We investigated temporal changes in ^{129}I activity and ^{129}I/^{127}I ratios in total suspended substance (SS) collected during December 2012– January 2014 at Haramachi on the Niida River. Particular attention is given to quantifying the monthly flux of particulate ^{129}I in the downstream reaches of the Niida River system.

6.2 Materials and Methods

The Niida River system in northeastern Fukushima Prefecture has a 78-km-long watershed with an area of 585 km^2 [5]. The upper watershed is located 30–40 km northwest of the FDNPP and is affected by relatively high levels of radioactive contamination, with ^{137}Cs inventory of >300 kBq m^{-2} (Fig. 6.1) [6]. In contrast,

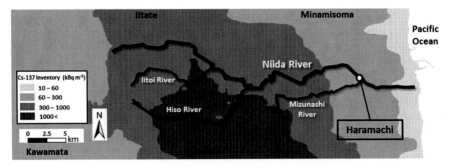

Fig. 6.1 Map showing the sampling site (Haramachi) located 5.5 km upstream from the Niida river mouth along with the spatial ^{137}Cs inventory from MEXT [6]. Total SS was collected continuously from December 2012 to January 2014 using a time-integrative SS sampler

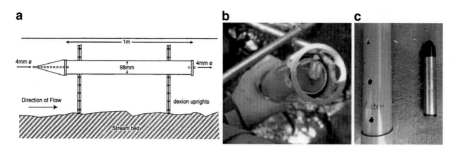

Fig. 6.2 (**a**) Schematic diagram of the time-integrative SS sampler. (**b**) Photographs of the tuner turbidity sensor, and (**c**) water level sensor for measuring discharge

the downstream reaches are characterized by low levels of contamination, with ^{137}Cs inventories of 10–300 kBq m^{-2} [6]. Mean annual precipitation is ∼1900 mm at Haramachi station in the downstream reaches and 1700–1800 mm at Iitate and Tsushima stations located near the upper watershed [7].

SS was collected continuously at Haramachi (5.5 km upstream from the river mouth) from December 2012 to January 2014 using a time-integrative SS sampler with an intake diameter of 4 mm (Fig. 6.2). This sampler was used successfully in a previous study of particulate ^{137}Cs flux in the Fukushima river system [8, 9]. Monthly turbidity (kg m^{-3}) and water discharge (m^3 month^{-1}) were calculated at sampling sites using data from a tuner turbidity sensor and water level sensor (Fig. 6.2). The flux of particulate ^{129}I at Haramachi can be estimated using the following function:

$$F\left(^{129}I\right) = A\left(^{129}I\right) \cdot T \cdot D \qquad (6.1)$$

where $F\left(^{129}I\right)$ is the flux of ^{129}I (Bq month^{-1}), $A\left(^{129}I\right)$ is the SS ^{129}I activity (Bq kg^{-1}), T is turbidity (kg m^{-3}), and D is water discharge (m^3 month^{-1}).

Samples for measurements of ^{129}I activity and ^{129}I/^{127}I ratios were prepared following Muramatsu et al. [10] and analyzed using an accelerator mass spectrometer (AMS) configured by Matsuzaki et al. [11]. Dried SS samples (∼0.5 g) were combusted with V_2O_5 at 1000 °C in a quartz tube for 30 min under a constant flow of pure O_2 and water vapor. Volatilized iodine was trapped in an organic alkaline solution. Stable iodine (^{127}I) in trap solutions was measured using an inductively coupled plasma-mass spectrometer (ICP–MS, Agilent 8800). After adding 2 mg of iodine carrier to the trap solution, the iodine was isolated and precipitated as AgI. The ^{129}I/^{127}I ratio of AgI targets was measured using an AMS system at the Micro Analysis Laboratory Tandem Accelerator (MALT), University of Tokyo. A terminal voltage of 3.47 MV and a charge state of 5+ were chosen for acceleration and detection. Measurement ratios were normalized against the Purdue-2 reference material, which has an ^{129}I/^{127}I ratio of 6.54×10^{-11} [12] and was obtained from the Purdue Rare Isotope Measurement Laboratory (PRIME Lab) at Purdue University. The overall precision of the system was better than 5 %, and the blank levels, which included the iodine carrier, were $2.2–4.9 \times 10^{-13}$ during all experimental procedures. The original $^{129}I/^{127}I$ ratio and ^{129}I activity in SS samples were calculated using the ^{127}I concentration obtained by ICP–MS and the $^{129}I/^{127}I$ ratio obtained by AMS.

6.3 Results and Discussion

6.3.1 Source of Particulate ^{129}I in the Niida River System

Table 6.1 lists the total dry weight, ^{129}I activity, and $^{129}I/^{127}I$ ratio for SS samples collected at Haramachi. As shown in Fig. 6.3, SS weights increased abruptly in March and April 2013, then increased continuously from May to October 2013. Based on meteorological data provided by the Japan Meteorological Agency, monthly precipitation was relatively high at Haramachi, Iitate, and Tsushima stations in April (140–160 mm), September (150–220 mm), and October (240–350 mm) [7]. Thus, the increased SS dry weights are thought to be related to higher-than-average precipitation in 2013.

SS ^{129}I activity and $^{129}I/^{127}I$ ratios in Niida River samples were 0.9–4.1 mBq kg^{-1} and $(2.5–4.4) \times 10^{-8}$, respectively. These values are 2–10 and 2–3 times larger, respectively, than the pre-accident level for ^{129}I activity of 0.42 mBq kg^{-1} and $^{129}I/^{127}I$ ratio of 1.6×10^{-8} at Fukushima before the FDNPP accident [13]. Higher SS ^{129}I activity and $^{129}I/^{127}I$ ratios were found in March, April, September,

Table 6.1 Suspend substance (SS) weight, ^{129}I activity, and $^{129}I/^{127}I$ ratios measured in samples collected at Haramachi in the downstream reaches of the Niida River system

Sampling month		Sampling period		Suspended substance	^{129}I activity	$^{129}I/^{127}I$ ratio
		Start date	End date	(g dry weight)	(mBq kg^{-1})	($\times 10^{-8}$)
2012	December	2012/12/6	2012/12/18	0.003	n.a.	n.a.
2013	January	2012/12/18	2013/1/22	0.35	n.a.	n.a.
	February	2013/1/22	2013/2/26	0.06	n.a.	n.a.
	March and April	2013/2/26	2013/4/18	6.70	2.06 ± 0.05	4.02 ± 0.11
	May	2013/4/18	2013/5/21	0.95	n.a.	n.a.
	June	2013/5/21	2013/6/18	2.16	0.92 ± 0.03	2.52 ± 0.09
	July	2013/6/18	2013/7/25	3.24	1.71 ± 0.06	3.24 ± 0.11
	August	2013/7/25	2013/8/22	5.51	2.15 ± 0.07	4.14 ± 0.14
	September	2013/8/22	2013/9/26	9.25	4.07 ± 0.20	4.39 ± 0.22
	October	2013/9/26	2013/10/30	7.47	2.86 ± 0.11	4.83 ± 0.20
	November	2013/10/30	2013/11/20	0.83	n.a.	n.a.
	December	2013/11/20	2013/12/23	0.26	n.a.	n.a.
2014	January	2013/12/23	2014/1/17	0.05	n.a.	n.a.

n.a. not analyzed

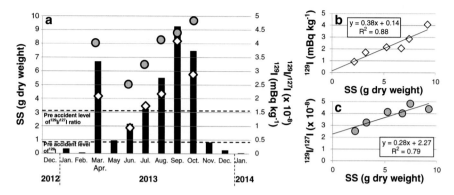

Fig. 6.3 (a) Temporal changes in monthly SS weight (*black bars*), ^{129}I activity (*open diamonds*), and ^{129}I/^{127}I ratio (*gray circles*) at Haramachi. (b) Correlations between ^{129}I activity and SS, and (C) the ^{129}I/^{127}I ratio and SS

and October 2013, when the SS weights were relatively high. The ^{129}I activity and ^{129}I/^{127}I ratios are strongly correlated with SS weight ($R^2 = 0.79$–0.88). As described in Sect. 6.2, the Niida River flows through highly contaminated areas in the upper watershed and medium–low contamination areas in the middle to lower reaches [6]. Therefore, it is possible that SS ^{129}I activity and ^{129}I/^{127}I ratios reflect the source of SS, i.e., either the more contaminated upper watershed or less contaminated downstream area. Further study is needed to clarify the differences in ^{129}I activity, ^{129}I/^{127}I ratios, and the ^{129}I inventory in soil between the upper watershed and downstream areas.

6.3.2 Flux of Particulate ^{129}I in the Niida River System

Table 6.2 lists the monthly flux of SS and associated ^{129}I at Haramachi. The SS flux and particulate ^{129}I are estimated to be 30–3200 ton month^{-1} and 0.1–9.0 kBq month^{-1} from March to October 2013, respectively. A higher ^{129}I flux of 7.6–9.0 kBq month^{-1} was recorded in September and October 2013, when high monthly precipitation (150–350 mm) was observed in the Niida River watershed [7]. As discussed in Sect. 6.3.1, particulate ^{129}I in September–October 2013 is considered to have originated mainly from the more highly contaminated upper watershed area. Therefore, a relatively high amount of particulate ^{129}I was transported from the upstream to downstream reaches of the Niida River by a rain event during September and October 2013. Further investigation is needed to better understand the ^{129}I flux in the river system.

Table 6.2 Monthly suspended substance flux and associated ^{129}I at Haramachi in the downstream reaches of the Niida River system

Analysis period		Analysis period		Suspended substance flux (ton month^{-1})	^{129}I activity (mBq kg^{-1})	^{129}I flux (kBq month^{-1})
		Start date	End date			
2013	March	2013/3/1	2013/4/1	45.1	2.06	0.09
	April	2013/4/1	2013/5/1	318	2.06	0.66
	May	2013/5/1	2013/6/1	30.3	n.a.	–
	June	2013/6/1	2013/7/1	110	0.92	0.10
	July	2013/7/1	2013/8/1	–	1.71	–
	August	2013/8/1	2013/9/1	–	2.15	–
	September	2013/9/1	2013/10/1	1860	4.07	7.58
	October	2013/10/1	2013/10/30	3160	2.86	9.03

n.a. not analyzed

6.4 Conclusions

Monthly SS ^{129}I activity and ^{129}I/^{127}I ratios measured from March to October 2013 in the downstream reaches of the Niida River system were 0.9–4.1 mBq kg^{-1} and $(2.5–4.4) \times 10^{-8}$, respectively. These values are strongly correlated with the total SS dry weight ($R^2 = 0.79$–0.88). The SS ^{129}I activity and ^{129}I/^{127}I ratio are considered to reflect the source of SS, i.e., the more contaminated upper watershed or the less contaminated downstream area. The particulate ^{129}I flux at Haramachi was estimated to be 7.6–9.0 kBq month^{-1} from September to October 2013. Relatively large amounts of particulate ^{129}I were transported by heavy rain from the upstream to downstream reaches of the Niida River over this period.

Acknowledgments We are grateful to the staff of MALT, The University of Tokyo, for technical assistance with ^{129}I measurements. This work was supported by JSPS KAKENHI Grant Numbers 24246156 and 24110006.

References

1. Hou XL, Povinec PP, Zhang LY, Shi K, Bidduiph D, Chang CC, Fan Y, Golser R, Hou Y, Jeskovsky M, Hull AJT, Liu Q, Lou M, Steier P, Zhou W (2013) Iodine-129 in seawater offshore Fukushima: distribution, inorganic speciation, sources, and budget. Environ Sci Technol 47:3091–3098
2. Chino M, Nakayama H, Nagai H, Terada H, Katata G, Yamazawa H (2011) Preliminary estimation of release amounts of ^{131}I and ^{137}Cs accidentally discharged from the Fukushima Daiichi Nuclear Power Plant into the atmosphere. J Nucl Sci Technol 48:1129–1134

3. Terada H, Katata G, Chino M, Naga H (2012) Atmospheric discharge and dispersion of radionuclides during the Fukushima Dai-ichi Nuclear Power Plant accident. Part II: verification of the source term and analysis of regional-scale atmospheric dispersion. J Environ Radioact 112:141–154
4. Morino Y, Ohara T, Nishizawa M (2011) Atmospheric behavior, deposition, and budget of radioactive materials from the Fukushima Daiichi nuclear power plant in March 2011. Geophys Res Lett 38:L00G11
5. Nagao S, Kanamori M, Ochiai S, Inoue M, Yamamoto M (2015) Migration behavior of ^{134}Cs and ^{137}Cs in the Niida River water in Fukushima Prefecture, Japan during 2011–2012. J Radioanal Nucl Chem 303:1617–1621
6. MEXT (2011) http://ramap.jmc.or.jp/map/eng. Accessed 20 May 2015
7. JMA (2015) http://www.data.jma.go.jp/obd/stats/etrn/index.php. Accessed 20 May 2015
8. Taniguchi K, Onda Y, Yoshimura K, Smith H, Blake W, Takahashi Y, Sakaguchi A, Yamamoto M, Yokoyama A (2014) Radiocesium transportation through river system in Fukushima. KEK Proc 2014–7:168–177
9. Yamashiki Y, Onda Y, Smith HG, Blake WH, Wakahara T, Igarashi Y, Matsuura Y, Yoshimura K (2014) Initial flux of sediment-associated radiocesium to the ocean from the largest river impacted by Fukushima Daiichi Nuclear Power Plant. Sci Rep 4:3714–3720
10. Muramatsu Y, Takada Y, Matsuzaki H, Yoshida S (2008) AMS analysis of ^{129}I in Japanese soil samples collected from background areas far from nuclear facilities. Quat Geochronol 3:291–297
11. Matsuzaki H, Muramatsu Y, Kato K, Yasumoto M, Nakano C (2007) Development of ^{129}I-AMS system at MALT and measurements of ^{129}I concentrations in several Japanese soils. Nucl Instrum Methods Phys Res Sect B 259:721–726
12. Sharma P, Elmore D, Miller T, Vogt S (1997) The ^{129}I AMS program at PRIME Lab. Nucl Instrum Methods Phys Res Sect B 123:347–351
13. Matsunaka T, Sasa K, Sueki K, Takahasi T, Matsumura M, Satou Y, Kitagawa J, Kinoshita N, Matsuzaki H (2015) Post-accident response of near-surface ^{129}I levels and ^{129}I/^{127}I ratios in areas close to the Fukushima Dai-ichi Nuclear Power Plant, Japan. Nucl Instrum Methods Phys Res Sect B 361:569–573

Part II
Decontamination and Radioactive Waste

Chapter 7
Safety Decontamination System for Combustion of Forestry Wastes

Hirohisa Yoshida, Hideki Ogawa, Kahori Yokota, Shio Arai, Shigemitsu Igei, and Ritsuko Nakamura

Abstract The safety decontamination system of the contaminated forestry wastes by combustion was developed. Under the laboratory scale test with 10 g of cedar bark, about 35 % of radiocesium in the contaminated bark flowed out as a gaseous state by the combustion above 500 °C. The developed system consisted of three modules, the smoke extraction apparatus by water, the combustion ash coagulate apparatus and the radiocesium filtration unit from the sewage water. The demonstration combustion tests were carried out in March 2012. Forestry wastes (6.3 kg), pine needles, Japanese cedar bark and sapwood chips including radiocesium were combusted at 550–700 °C. The exhaust smoke was washed by the jet stream of water, the sewage water included small amount of soot and the radiocesium concentration of sewage water without soot was 50 Bq/kg. After the filtration of 550 L of sewage water by the radiocesium absorption filter consisting of wool dyed by Prussian blue, the radiocesium concentration decreased less than 0.2 Bq/kg. The filtrated water was recyclable in this system. No gaseous radiocesium was detected in the exhausted air from this system during the decontamination of forestry wastes. The combustion ash (140 g), consisting of cesium oxide alloy including various metal ions, was collected and packed under the reduced pressure automatically.

Keywords Forestry wastes • Combustion • Smoke extraction • Radiocesium • Decontamination • Sewage water • Cedar • Bark • Pine needles • Ash

H. Yoshida (✉) • K. Yokota • S. Arai • S. Igei • R. Nakamura
Graduate School of Urban Environmental Science, Tokyo Metropolitan University,
Hachioji, Tokyo, Japan
e-mail: yoshida-hirohisa@tmu.ac.jp

H. Ogawa
Graduate School of Urban Environmental Science, Tokyo Metropolitan University,
Hachioji, Tokyo, Japan

Fukushima Forestry Research Center, Fukushima-shi, Fukushima, Japan

© The Author(s) 2016
T. Takahashi (ed.), *Radiological Issues for Fukushima's Revitalized Future*,
DOI 10.1007/978-4-431-55848-4_7

7.1 Introduction

Although 4 years have passed after the accident of Tokyo Electric Power Co. (TEPCO) Fukushima Daiichi Nuclear Power Station (FDPS), a huge amount of radionuclide remains in the forest at Fukushima [1]. The main components of radionuclide from FDPS disaster, cesium134 (^{134}Cs) and cesium137 (^{137}Cs), became the environmental contamination problem and disturbed both agricultural and forestry works in Fukushima [2, 3]. Most radiocesium fallout on the ground tightly bound with soil and formed the complex with clay minerals [4], the ionic ^{134}Cs and ^{137}Cs were a few in the environment [5, 6].

Fukushima Daiichi Nuclear Power Station accident induced the serious situation to revive the forestry works in Fukushima, especially in Abukuma forest area located at west from FDPS with heavy radioactive contamination. The regular maintenance of forest became difficult due to the high air dose environment, the low demand of woods and the forestry wastes including radionuclide. The fallout radiocesium (^{134}Cs and ^{137}Cs) existed on bark and leaves of trees [7–9], and the Japanese cedar woods absorbed radiocesium through bark and leaves in 2011 and 2012 [10, 11].

In Fukushima prefecture, about 970,000 ha of forest is the subject area for the periodic thinning from 2013 to 2030 in the current plan. For 1000 ha of the thinning area in forest, about 10 % of woods corresponding to 150,000 m^3 of Japanese cedar will be cut down. The thinning of 1000 ha of forest will produce about 30,000 m^3 of twigs including leaves and about 15,000 m^3 of barks. The total forestry wastes produced by the periodic thinning is estimated at 3,445,000,000 m^3 for each year. Before FDPS accident, the forestry wastes were used as composts for forestry and agriculture uses. The forestry wastes including radiocesium or other radionuclide have no way for utilization. These forestry wastes are disposable by the combustion or the fermentation as the energy resource.

One of the possible ways of forestry waste disposal is the biomass power generation. Ministry of the Environment planed the construction of two biomass power generations with 12 MW at Samekawa village and Hanawa town in 2013. The benefit of biomass power generation is the effective utilization of forestry waste and the power generation; on the other hand, the release of gaseous radionuclide and the mass transportation of biomass contaminated with radionuclide from the forest to the power plant are the risk of biomass power generation in Fukushima at present. The construction of biomass power generation (10 MW scale) in Hanawa town was canceled in October 2013 by the local residents' campaign based on the unclearness of these risks [12]. However, Fukushima prefecture decided to construct again the biomass power generation in Hanawa town in May 2015 [13]. The insecurities of large-scale biomass power station are the leak of radionuclide from the discharge gas during the burning of forestry wastes, the fly out risk of the concentrated ash and the radiation exposure of workers in the biomass power plant.

In this study, the risk of radiocesium released in the environment during the combustion of forestry waste was estimated by the laboratory-scale combustion experiment. The demonstration combustion test of forestry wastes was carried to evaluate the possibility of safety combustion method. The safety combustion system designed based on the results of laboratory-scale experiments was developed for the decontamination of forestry wastes.

7.2   Experimental

7.2.1   Samples

Samples used in this study were Japanese cedar bark (from 5000 to 21,000 Bq/kg), cedar leaves (from 5000 to 20,000 Bq/kg), cedar sapwood (less than 20 Bq/kg), cedar heartwood (less than 10 Bq/kg) and pine needles (from 10,000 to 25,000 Bq/kg) collected at Fukushima Forestry Research Center in autumn 2011. After drying at 60 °C for 3 days, samples were used for the experiments. The water content of samples evaluated from the mass loss after drying was 28.6 % for cedar bark, 56.8 % for cedar leaves, 61.8 % for sapwood, 54.3 % for heartwood and 10 % for pine needle based on the wet mass.

7.2.2   Combustion Test

Laboratory-scale combustion tests of radiocesium-contaminated Japanese cedar bark (20,000 Bq/kg) were carried out using the combustion instrument made of glass shown in Fig. 7.1. About 10 g of cedar bark in alumina crucible (A) was put in the quartz sample tube (B) and combusted at 300–900 °C under air flow condition. The combustion gas was cooled at the glass tube (C) by ice/water bath and was collected

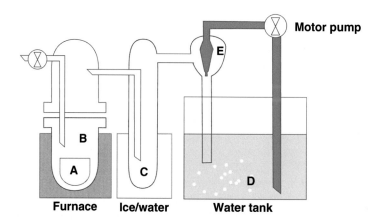

Fig. 7.1 Schematic drowning of combustion instrument for laboratory scale test

in 2 L of water sink (D) using an aspirator (E). Before and after the combustion, the radioactivity of sample and water in sink D were determined. The mass loss by the combustion was also determined.

Demonstration tests of forest waste were carried out at Fukushima Forestry Research Center in March 2012. Two times demonstration tests using 2.3 kg of pine needles, 3.6 kg of Japanese cedar bark and 0.4 kg of cedar sapwood were carried out by the combustion system described later.

7.2.3   Measurements

Radioactivity of samples was measured by the Germanium semiconductor detector (SEG-EMS: SEIKO EG&G Inc., Japan) with 100 mL of U8 container at 10,000–40,000 s of accumulation time to obtain 1–2 Bq/kg of the lower limit. The radiation decay was compensated at March 1, 2012.

Thermogravimetry measurements of forest wastes, Japanese cedar sapwood chip, cedar bark, cedar leaves and pine needles were carried out by TG/DTA 7200 (Hitachi High-Tech Science Co., Japan) under air and nitrogen gas flow atmospheres. About 3 mg of samples were heated at 20 K/min from room temperature to 900 °C. The evolved gas during combustion was analyzed by the online Fourier Transfer Infrared Spectroscopy FTIR 650 (JASCO, Japan) connected with TG/DTA [14].

Transmission electron microscopic observation of ash was performed by JEM 3200FS (JEOL, Japan) operated at 300 kV. The ash sample was dispersed on the copper glide and coated with carbon. The energy-dispersive X-ray spectroscopic analysis (EDX) was carried out for ash sample. Powder X-ray diffraction of ash was observed by Rint TTR III (Rigaku Co., Japan) operated at 50 kV and 300 mA with $2\theta/\theta$ scanning at 2°/min.

7.3   Results and Discussion

7.3.1   Laboratory-Scale Tests

Mass loss curves of Japanese cider sapwood chip, bark, leaves and pine needles obtained by TG/DTA were shown in Fig. 7.2 in air flow (left) and nitrogen gas flow (right) atmospheres. The mass loss occurred in three steps below 200 °C, 200–400 °C and above 400 °C in both conditions; however, the third mass loss in air atmosphere occurred in the narrow temperature range; in contrast, the third step in nitrogen atmosphere occurred in the wide temperature range. The first mass loss below 200 °C was caused by the evaporation of water and volatile components without the combustion of main components of forestry waste, such as cellulose,

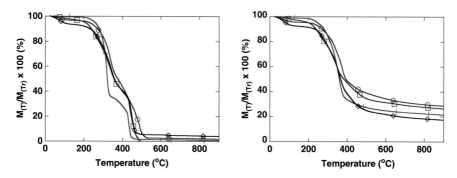

Fig. 7.2 Mass change of Japanese cedar sapwood chip (*cross*), cedar bark (*circle*), cedar leaves (*square*) and pine needles (*rhombus*) under air (*left*) and nitrogen gas (*right*) gas atmosphere

lignin and hemicellulose. The main component of mass loss below 150 °C was water from the evolved gas analysis. The second mass loss occurred above 250 °C and levelled off below 400 °C for both conditions; these mass losses were caused by thermal decomposition of cellulose and hemicellulose. The major components evolved in this stage were CO_2 and CH_4 in air atmosphere and CO and CH_3OH in nitrogen atmosphere. The third mass loss stage was caused by the dehydration of samples; the major components in this stage were CO_2, CH_4 and phenolic derivatives for air atmosphere and CO and H_2 for nitrogen atmosphere which were almost similar to the compounds obtained by the pyrolysis of wood [15].

The residues of each sample at 800 °C and 890 °C were 0.8 % and 0.6 % for Japanese cider sapwood chip, 1.9 % and 1.7 % for bark, 0.5 % and 0.4 % for leaves, 4.2 % and 4.0 % for pine needles in air atmosphere and 22.4 % and 21.4 % for sapwood chip, 29.8 % and 28.9 % for bark, 27.7 % and 26.6 % for leaves and 18.2 % and 17.3 % for pine needles in nitrogen atmosphere, respectively. The residues obtained in air and nitrogen atmospheres were inorganic materials and char, respectively, by XRD analysis of residues.

The mass loss and the radiocesium loss, normalized by the initial mass and the initial concentration of ^{134}Cs and ^{137}Cs, during the combustion of cedar bark at various temperatures were shown in Fig. 7.3. Under the laboratory scale combustion test, the mass loss at the combustion temperature below 500 °C was less than 90 %, which was lower than the mass loss evaluated by TG measurement under air flow condition. However, the mass loss behavior occurred in three steps, which was consistent with TG behavior shown in Fig. 7.2.

For the combustion at 300 °C, most radiocesium (98 %) remained in the residues, which was 46 % of the initial mass; however, radiocesium in the residues decreased with the increase of combustion temperature. About 15–25 % of radiocesium evaporated by the combustion at temperature from 350 to 450 °C and about 30–35 % of radiocesium released by the combustion at temperature above 500 °C. After seven combustion tests at temperatures from 300 to 900 °C, the radiocesium concentration (^{134}Cs $+^{137}$Cs) of trapped water was 153 Bq/L, which corresponded

Fig. 7.3 Mass loss (*open circle*) and radiocesium loss of ^{134}Cs (*closed circle*) and ^{137}Cs (*closed square*) at various combustion temperature

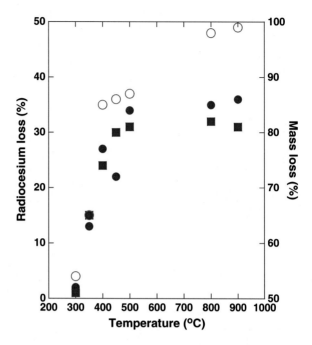

to 98 % of radiocesium evaporated during the combustion tests. These results indicated radiocesium in forestry wastes vaporized by the combustion, and the evaporated radiocesium was trapped efficiently by water.

7.3.2 Development of Decontamination System for Forestry Waste

In Fukushima, the non-industrial wastes including contaminated materials are burned at about 900 °C in the waste incineration plant equipped with the smoke extraction apparatus such as the desulfurization equipment and the bag filter for fly ash. The desulfurization of exhaust gas is processed generally by the absorption of sulfurous acid gas (SO_2) with aqueous dispersion of limestone ($CaCO_3$), which has the possibility to trap the evaporated radiocesium; however, the desulfurization equipment is not designed for the radiocesium absorption. Fly ashes are filtrated by the bag filter made of poly(ethylene tetrafluoride) with 0.4 μm of mesh size.

The safety decontamination system for forestry wastes was designed as shown in Fig. 7.4 [16]. The design criteria of this system were the compact size and the flexible attachment to connect with the existing small combustion furnace (100 kW scale) in sawmill and wood market yard. The developed system consisted of three modules, the smoke extraction apparatus by water (I), the combustion ash coagulate apparatus (II) and the radiocesium filtration unit from the sewage

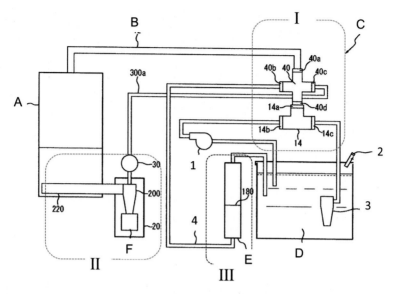

Fig. 7.4 Design of combustion system consisted of furnace (**A**), smoke tube (**B**), water scrubber absorbed smoke (**C**), water tank (**D**), sewage water filtrate column (**E**) and ash collecting apparatus (**F**)

water (III). The furnace (A) connected with the smoke extraction apparatus by jet stream of water (C) through the smoke tube (B), water supplied from the water tank (D), the combustion ash was collected by the ash enclosure apparatus (F) and the radiocesium in the sewage water was filtrated by the absorption column unit (E).

The woodstove model 1630CB (Morsφ, Finland) connected with the water scrubber (14) through the smoke tube and three channel connector (40). About 600 L of water stored in water tank (D) pumped up and compressed by motor (1), and water jet (2 L/s) flowed into the scrubber from 14b and flowed out from 14c induced the reduced pressure at 14a. The smoke from the woodstove was absorbed in the scrubber from 14a and mixed with water jet in the scrubber. The mixed smoke and water flowed in the water tank through the wet cyclone (3) where the radiocesium in smoke dissolved in water and the excess pressured air flowed out through the air filter made of wool dyed by Prussian blue (2).

After the combustion, the water-extracted smoke was decontaminated by radiocesium absorption filter made of wool dyed by Prussian blue (18) through the column (E) absorbed by the scrubber through three mouth connector (40b). The combustion ash including radiocesium was collected and separated particles by air cyclone (200), and the separated ash particles were packed automatically under the reduced pressure by the ash enclosure (F: ATA Co, Japan) connected with the scrubber through three mouth connector (40c).

The outlook of combustion system was shown in Fig. 7.5. The smoke extraction apparatus (1) consisted of the water tank (A), the water motor (B), the scrubber (C)

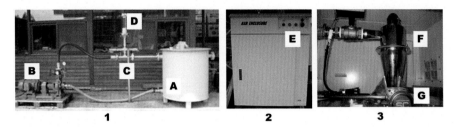

Fig. 7.5 Photographs of the smoke extraction system (*1*), the ash enclosure instrument (*2*) and its inside view (*3*)

and the three mouth connection to furnace (D) and the front view of ash enclosure (2) with control panel (E) and its inside consisted of air cyclone (F) and ash packing instrument (G) were shown in Fig. 7.5.

7.3.3 Demonstration Test of Combustion for Forestry Wastes

The decontamination experiments were carried out at Fukushima Forestry Research Center in March 2014. Two combustion tests using 2.3 kg of pine needles with 1.0 kg of cedar bark and 2.5 kg cedar bark with 0.4 kg of cedar sapwood were burned with slow temperature increment from 400 to 700 °C for 1 h. The mass and radiocesimu concentration in forestry wastes used for the combustion tests were listed on Table 7.1. About 550 L of water was used for the smoke extraction. During the demonstration tests, the surrounding temperature was about 4–6 °C; the initial temperature of water was about 5 °C and became 45 °C after the combustion tests. After the combustion test, about 5 L of water was lost by evaporation. From these values, the heat transfer from smoke to water was 2.9 kWh, which corresponded to 200 kWh scale combustion furnace, such as wood chip boiler and incinerator. No radiocesium was detected (the lower limit of 0.2 Bq/kg) in the air filter where the exhausted steam passed from this system during the decontamination.

Immediately after the combustion tests, the clear water changed the black soot-suspended water with pH 7.2. The sewage water absorbed the acidic compound caused by the decomposition of wood components and the soot. The suspended soot in sewage absorbed on the surface of water tank overnight; the optical clear and pale yellow sewage water was obtained. The smoke tube with 5 m length took off from the system and separated to pieces; the soot absorbed in smoke tube was washed by 100 L of water. The sewage water (50 Bq/L, about 545 L) and the soot-washed water (soot suspended water, 60 Bq/L, about 100 L) were decontaminated by the filter system with 300 g of Prussian blue–dyed wool filter. After 1 h filtration, no radiocesium was detected by Ge semiconductor detector with 0.2 Bq/L of the lower limit. The decontaminated water was recyclable in this system.

Table 7.1 Mass and radiocesium concentration of materials used for the combustion test and obtained after the combustion

Raw materials of combustion			Collected materials after combustion		
	$^{134}Cs + ^{137}Cs$	Mass		$^{134}Cs + ^{137}Cs$	Mass
Materials	Bq/kg	kg	Materials	Bq/kg	kg
Pine needle[a]	22,000	2.3	Ash	440,000	0.14
Cedar bark[a]	10,800	1.0	Soot in water[b]	290,000	>0.0015
Cedar bark[c]	15,000	2.5	Soot in flue[d]	60	100
Sapwood[c]	50	0.4	Sewage water[e]	50	545

[a]Materials used for the first combustion test
[b]Soot suspended in sewage water absorbed on water tank, collected from the water tank surface
[c]Materials used for the second combustion test
[d]Soot absorbed in smoke tube washed by 100 L of water
[e]Sewage water after filtration of suspended soot

The ash was collected by the enclosure under the reduced pressure and was packed and sealed automatically in the antistatic polyethylene bag under vacuum. After sealing the ash, the package of ash was easily handled without further contamination. The mass and radiocesium concentration of ash, sewage water, separated soot and washed soot from the smoke tube were shown in Table 7.1. The ratio of mass of ash against the raw materials was 22.2 %, which indicated that the combustion proceeded in imperfect condition. The soot suspended in sewage water and absorbed in smoke tube also suggested the imperfect combustion. The collected ash included white powder and black charcoal particles caused by the imperfect combustion.

The mass balance of radiocesium before and after the combustion was evaluated. The initial amount of radicesium in the raw materials was 98,920 Bq; the amount of radiocesium in the collected materials was 95,285 Bq. About 95 % of radiocesium was recovered by this combustion system. The difference between total amount of radiocesium before and after the combustion was caused by the uncertainty of initial concentration due to the inhomogeneity of radiocesium distribution in pine needles and bark, and the radiocesium remained as soot in the woodstove, smoke tube and water tank.

The transmission electron microscopic observation and EDX spectrum of white ash were shown in Fig. 7.6. The main component of white ash was calcium oxide from X-ray diffraction analysis. EDX spectrum analysis indicated the existence of boron (0.187 keV, BKα), magnesium (1.25 keV, MgKα), silica (1.74 keV, SiKα), phosphor (2.01 keV, PKα), manganese (5.89 keV, MnKα), iron (6.38 keV, FeKα), copper (8.0 keV, CuKα), zinc (8.63 keV, ZnKα) and strontium (14.1 keV, SrKα) addition to calcium (3.69 keV, CaKα). Mg, P, Mn, Zn and Ca were the base element of plants, and B, Si, Fe, Cu and Sr came from the soil contamination of plants. The existence of radiocesium was not clear due to the low concentration in ash, because the estimated radiocesium concentration was about 12 ppb evaluated from 440,000 Bq/kg. The main component of calcium oxide formed the solid solution

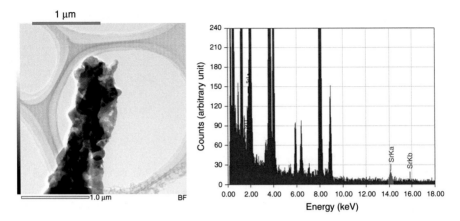

Fig. 7.6 Transmutation electron microscopic image (*left*) and EDX spectrum (*right*) of ash

with these elements. The water-soluble radiocesium from 5 g of ash was extracted by 4 L of water stirring for 5 h at room temperature. After the filtration of insoluble ash, the filtered water contained 260 Bq/L of radiocesium, which indicated 47 % of radiocesium in ash was water soluble. This result suggested that the storage of the ash from forestry wastes should be done with the extreme care not to leak the water-soluble radiocesium to the environment.

7.4 Conclusion

The mass balance of radiocesium in forestry wastes by combustion was evaluated with the laboratory-scale combustion tests and the demonstration combustion tests. About 30–35 % of radiocesium in cedar bark evaporated by the combustion under air flow condition at temperature above 500 °C. Most evaporated radiocesium was trapped efficiently by water.

Based on the laboratory combustion test, the decontamination system for the combustion of forestry wastes was developed. The decontamination system consisted of three modules, the smoke extraction apparatus by water, the combustion ash coagulate apparatus and the radiocesium filtration unit from the sewage water.

The demonstration combustion test for forestry wastes using the developed decontamination system was carried out. No gaseous radiocesium was detected in the exhausted air from this system during the decontamination of forestry wastes. About 80 % of mass of forestry waste decreased by the combustion, and 95 % of radiocesium was collected safely by the developed system. The ash caused by the combustion of forestry waste contained about 45 % of water-soluble radiocesium.

References

1. MECSST; Ministry of Education, Culture, Sports, Science and Technology (2011) 6th air dose monitoring by aircraft. http://radioactivity.nsr.go.jp/ja/contents/7000/6749/24/191_258_0301_18.pdf
2. Nakanishi T, Tanoi K (eds) (2013) Agricultural implications of the Fukushima nuclear accident. Springer, Tokyo
3. Yoshida H (2012) Contamination analysis of radionuclide in soils and plants caused by Fukushima nuclear power plant accident. Chem Eng 6:250–257
4. Kikawada Y, Hirose M, Tsukamoto A, Nakamachi K, Oi T, Honda T, Takahashi H, Hirose K (2015) Mobility of radioactive cesium in soil originated from the Fukushima Daiichi nuclear disaster; application of extraction experiments. J Radioanal Nucl Chem 304(1):27–31
5. Matsuda N, Mikami S, Shimoura S, Takahashi J, Nakano M, Shimada K, Uno K, Hagiwara S, Saito K (2015) Depth profiles of radioactive cesium in soil using a scraper plat over a wide area surrounding the Fukushima Dai-Ichi nuclear power plant, Japan. J Environ Radioact 139:427–434
6. Yoshida H (2014) Transportation and restraint of radiocesium in forest near the populated area. Green Power 2:14–15
7. Kuroda K, Kagawa A, Tonosaki M (2013) Radiocesium concentrations in the bark, sapwood and heartwood of three tree species collected at Fukushima forests half a year after the Fukushima Dai-ichi nuclear accident. J Environ Radioact 122:37–42
8. Yoshihara T, Matsumura H, Hashida S, Nagaoka T (2013) Radiocesium contaminations of 20 wood species and the corresponding gamma-ray dose rates around the canopies at 5 months after the Fukushima Nuclear Power Plant accident. J Environ Radioact 115:60–68
9. Ogawa H, Ito H, Yokota K, Arai S, Masumoto K, Yoshida H (2015) Radiocesium distribution and changes of the wood in Sugi and Hinoki caused by the accident of Fukushima Daiichi Nuclear Power Plant. In: Proceedings of the 16th workshop environmental radioactivity, KEK, Japan, vol 2015-8, pp 302–309. http://ccdb5fs.kek.jp/tiff/2015/1525/1525004.pdf
10. Ogawa H, Itou H, Murakami K, Kumata A, Hirano Y, Yokota K, Yoshida H (2014) Radiocesium distribution in Japanese Cedar trees at one year after Fukushima Dai-ichi nuclear power plant accident. In: Proceedings of the 15th workshop on environmental radioactivity, KEK, Japan, vol 2014-7, pp 252–260. http://ccdb5fs.kek.jp/tiff/2014/1425/1425007.pdf
11. Mahara Y, Ohta T, Ogawa H, Kumata A (2014) Atmospheric direct uptake and long-term fate of radiocesium in trees after the Fukushima nuclear accident. Sci Rep 4:7121–7127
12. Internet news, Fukushima Minyu, October 11, 2013. http://www.minyu-net.com/osusume/daisinsai/serial/131011/news3.html
13. Internet news, Fukushima Minpo, May 5, 2015. http://www.minpo.jp/pub/topics/jishin2011/2013/02/post_6181.html
14. Ozawa T (2004) Thermal analysis. In: Sorai M (ed) Comprehensive handbook of calorimetry and thermal analysis. JSCTA, Wiley, Chichester
15. Tsuge A, Outani H, Watanabe C (eds) (2006) Pyrolysis-GCMS of highpolymers, Techno-system, Tokyo
16. Tokyo Metropolitan University, Closed smoke extract system, Japanese Patent 2013-224831

Chapter 8
Remediation Technology For Cesium Using Microbubbled Water Containing Sodium Silicate

Yoshikatsu Ueda, Yomei Tokuda, and Hiroshi Goto

Abstract Remediation of materials contaminated with a radioactive material such as ^{137}Cs is important for public health and environmental concerns. Here, we report the effectiveness of aqueous sodium metasilicate (SMC) prepared using a microbubble crushing process for the removal of radioactive ^{137}Cs from contaminated materials. We have already reported that almost 80 % [137]Cs removal was achieved for a nonwoven cloth sample in which multiple washings using low SMC concentrations were effective. In addition, the volume of the waste solution can be reduced by neutralizing the SMC and using gelation to remove the radioactive material. We also attempt to clarify the mechanism of SMC operation by measuring its electrical properties. Decontamination is shown to be more efficient with SMC than with sodium hydroxide, even for washing granule conglomerates.

Keywords Radioactive cesium • Microbubble • Sodium metasilicate • Cs-137

8.1 Introduction

The accident at the Fukushima Daiichi nuclear power plant in 2011 following the Great East Japan Earthquake resulted in the dispersal of radioactive Cs into the environment and the contamination of an extensive area of soil. Various decontamination methods have been developed and applied in Fukushima Prefecture [1–4], but optimum methods for remediation of materials under specific contamination conditions still need to be developed. Radioactive Cs adsorbs on soil particles through ion exchange with potassium [5, 6]. Our research focused on using aqueous

Y. Ueda (✉)
Research Institute for Sustainable Humanosphere, Kyoto University, Gokasho, Uji,
Kyoto 611-0011, Japan
e-mail: yueda@rish.kyoto-u.ac.jp

Y. Tokuda
Institute for Chemical Research, Kyoto University, Gokasho, Uji, Kyoto 611-0011, Japan
e-mail: tokuda@noncry.kuicr.kyoto-u.ac.jp

H. Goto
Kureha Trading Co., Ltd., 1-2-10, Horidomecho, Nihonbashi, Chuo-ku, Tokyo 103-0012, Japan

© The Author(s) 2016
T. Takahashi (ed.), *Radiological Issues for Fukushima's Revitalized Future*,
DOI 10.1007/978-4-431-55848-4_8

Fig. 8.1 Photographs of aqueous sodium metasilicate irradiated by a green laser after storage for one-half year. The metasilicate was dissolved in water (**a**) without crushing, and (**b**) with crushing [1]. The *black* object in the lower right corner is the green laser

sodium metasilicate as a new detergent and possible decontamination agent. We examined the mechanism and principle of the change in the chemical reaction characteristics of aqueous sodium metasilicate after microbubble and ultrasonic treatments. We found that, because aqueous sodium metasilicate is not a surfactant, it has a low environmental load and does not exhibit foaming characteristics. It can be used with hard, soft, or sea water and is a "peeling detergent" in terms of its cleaning activity, which differentiates it from dissolving detergents such as organic solvents. As a peeling detergent, aqueous sodium metasilicate is suitable for foaming, jet streaming, high-pressure ultrasonic wave, and spray cleaning. A microbubble crushing process [7, 8] can also be used to suppress precipitation during long-term storage [9] (see Fig. 8.1).

Coprecipitation can be used to decrease the volume of the waste solution produced because sodium metasilicate turns into a gel that traps Cs when neutralized with acid [10]. Coprecipitation followed by gel formation can be used to separate the ^{137}Cs gel from the soil, yielding a decontaminated waste product.

8.2 Experimental

8.2.1 Preparation of Aqueous Sodium Metasilicate

A 0.47 mol/kg solution of sodium metasilicate nonahydrate ($Na_2O_3Si \cdot 9H_2O$) was prepared by dissolution in filtered water (manufactured using G-20B; Organo Corporation). A microbubble generator (capable of generating >20,000 microbubbles/mL) was manufactured by Kyowa Engineering and used to interfuse microbubbles into the solution. Ultrasonic irradiation (40 kHz; ultrasonic wave generator UT-1204U and ultrasonic transducer UI-12R3; Sharp Corporation) was employed to crush the bubbles, and the resulting solution, aqueous sodium metasilicate prepared with a microbubble crushing process (abbreviated here as SMC), was used as the nonsurfactant aqueous detergent. Purified water and aqueous sodium hydroxide with the same pH as SMC (pH = 13.1) were prepared for comparison. To analyze the structure of the sodium metasilicate in SMC and understand the dissolution stability, we measured the electrical conductivity and dissolved oxygen content every 5 min during the microbubble crushing process. We also compared the change in conductivity of SMC after 6 h.

8.2.2 Cleaning Method

Depending on the individual diameters and buoyancy, microbubbles only remain in water for several minutes. They can remain in an aqueous environment for longer times, however, if an ultrasound treatment is used to reduce their size [11]. An ultrasound pretreatment of aqueous sodium metasilicate to form SMC and reduce the bubble size does not decrease the eduction rate of aqueous sodium metasilicate and offers the possibility of sustained cleansing. In this experiment, we conducted a cleaning test in a standing solution so that we could focus on the chemical cleansing effects due to the synergy between aqueous sodium metasilicate and microbubbles as opposed to focusing on the physical cleansing effects [1].

Granule conglomerates and nonwoven cloths were used as the materials to be cleaned. The ~150-g granule conglomerate samples were collected from soil in a hot spot (~500,000 Bq/kg) at the Fukushima Agricultural Technology Centre. These pieces of nonwoven cloth (made of polypropylene) were used in farm work at farms exposed to the fallout from the nuclear accident in Fukushima Prefecture. The cloths had an average weight of 2.65 g, and the average amount of ^{137}Cs exhibited approximately 1633 Bq/sample (616,226 Bq/kg) of radioactivity.

The granule conglomerate and nonwoven cloth samples were immersed in 100 mL of several SMC concentrations for 20 h. Twelve granule conglomerate samples were prepared. These samples were divided into four groups of three samples. One group was washed in NaOH, one in normal sodium metasilicate, and the remaining two were washed in 10- and 100-wt% solutions of SMC. The 28 cloth samples were divided into 4 groups of 7 samples, and 1 group was washed in water, while the other 3 groups were washed in 1-, 10-, and 100-wt% solutions of SMC. To examine the effects of multiple washings, samples were tested after 6 h of immersion before the second and third washes. After these immersion tests, the samples were dried at 40 °C for 40 h until they were free of moisture [2].

To reduce the volume of waste solution, hydrochloric acid (HCl) was added to neutralize the detergent solutions after washing, which resulted in gelation. The gel and clear supernatant were then separated by filtration, and the radiation intensities (counts per second, cps) of the gel and supernatant were measured. We tried to remove ^{137}Cs from SMC solutions by neutralization. We prepared three samples of 10 and 100 wt% SMC solutions, and then neutralized and filtered them. We measured the radiation intensity and compared the results of the two remediation methods.

8.2.3   Radiation Measurement

The background radiation intensity of the nonwoven cloth specimens was measured using the germanium semiconductor detector at the Radioisotope Research Center, Kyoto University. The main unit of the detector is made from high-purity germanium (GMX-18200-S; EG&G Ortec), with a germanium crystal that is 102 cm^3 and a relative efficiency of 22.3 % (efficiency ratio of a $3'' \times 3''$ NaI (TI) (76×76 mm) crystal relative to that of the ^{137}Cs 662 keV gamma ray). The entrance window is a 0.5-mm-thick beryllium plate, which allows the detection of X-rays 3 keV or higher in energy, as well as high-energy gamma rays. The energy resolution was 0.54 keV for ^{55}Fe 5.9-keV (Mn Kα) X-rays and 1.8 keV for ^{60}Co 133-MeV gamma rays. A special container (100 mL) was used to analyze the nonwoven cloth specimens. The removal ratio was defined as the ratio of the radiation intensity from the sample after and before immersion, as shown in Eq. 8.1.

$$\text{Removal ratio } [\%] = \frac{\text{After immersion [cps]}}{\text{Before immersion [cps]}} \times 100 \qquad (8.1)$$

8.3   Results

8.3.1   Electrical Properties of SMC

Figure 8.2 shows the measured electrical conductivity and dissolved oxygen content during the SMC-making process. The electrical conductivity of each sample was measured every 5 min. In Fig. 8.3, we show real-time electrical conductivity data sampled 6 h after the data in Fig. 8.2. The results show that SMC is almost stabilized after 10 min, but compared with the real-time data, 1 h of microbubble crushing was required to stabilize the electrical conductivity.

8.3.2   Washing the Granule Conglomerate

The removal ratios of ^{137}Cs from the granule conglomerate using SMC (10 and 100 wt%), NaOH, and aqueous sodium hydroxide (Fig. 8.4) show that, while the decontamination efficiency of SMC was low, it was still more than 10 times higher than using sodium hydroxide at the same pH. Furthermore, the higher efficiency of SMC compared with sodium metasilicate without the microbubble treatment confirms the effectiveness of SMC. One of the characteristics of SMC is that the concentration of dissolved oxygen does not increase after aeration. Hence, a large number of adsorption pits (sites) are probably present, similar to zeolites. Thus, cesium ions present in water will be incorporated into the silicate. In addition, the

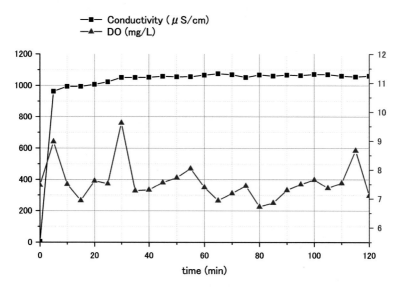

Fig. 8.2 Electrical conductivity and dissolved oxygen content during the microbubble crushing process

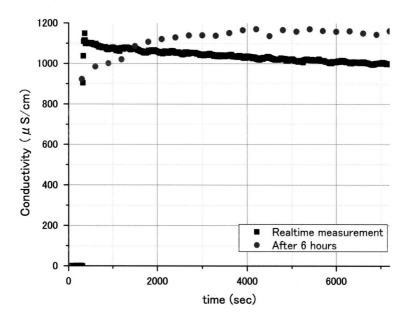

Fig. 8.3 Plot of conductivity versus time during the microbubble crushing process

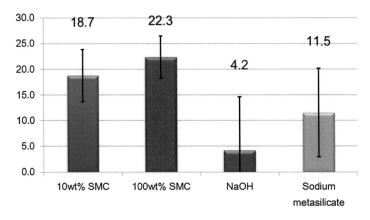

Fig. 8.4 Average removal ratios of ^{137}Cs from the granule conglomerate samples using 10- and 100-wt% SMC, NaOH, and sodium metasilicate

Table 8.1 ^{137}Cs removal ratios of using purified water and SMC for the nonwoven cloth samples*

Washing	Pure water	1-wt% SMC	10-wt% SMC	100-wt% SMC
First	0	44.5	59.5	77.5
Second	–	34.5	27.5	51.4
Third	–	20.6	–	–

*Pure water was used for the first washing, whereas 10- and 100-wt% SMCs were used for the second washing

decontamination efficiency increased because the alkaline solutions dissolved the granule conglomerate.

8.3.3 Washing the Nonwoven Cloth

The cleaning performance of SMC at various concentrations is compared in Table 8.1, which lists the average removal ratios for multiple washings with SMC concentrations of 1, 10, and 100 wt% compared with pure water [2].

As expected, almost no ^{137}Cs was removed by washing with pure water, but significant decontamination occurred as soon as SMC was introduced. Since the nonwoven cloth samples were previously used in agricultural fields, they also contained traces of fertilizers and other organic materials. As mentioned previously, sodium metasilicate detergent is used for washing because it has a capacity for breaking down organic materials via saponification, which is a result of its alkaline nature. We anticipate that we will be able to increase the efficiency of this detergent using microbubbles and ultrasound treatments. Organic components such as sebum and oil that are contaminated with ^{137}Cs are likely to be eluted from the nonwoven cloth owing to the alkalinity of SMC. The data in Table 8.1 also show that even with 1 wt% SMC, the removal ratio is 60 % of that obtained with 100 wt% SMC after the first washing. This result shows that the amount of detergent used for remediation can be reduced to a low concentration.

In the case of multiple washings, the background radiation intensity decreased considerably after the third washing using a highly concentrated (10 or 100 wt%) SMC solution. Therefore, a 1 wt% SMC solution was used for further analysis. The removal ratios after each wash with the 1 wt% solution, listed in Table 8.1, were 64 % for the second wash and 71 % for the third wash, indicating that even at a concentration of 1 wt%, significant decontamination of nonwoven cloth materials can be achieved after multiple washings. Though there was only a slight decrease in the removal ratios compared with higher wt% SMC solutions, we believe that using 1 wt% SMC is an effective washing method for even higher levels of contaminants because the need for further decontamination of materials in the Fukushima Prefecture still exists.

8.3.4 Remediation of ^{137}Cs by Neutralizing SMC

Reduction in the waste solution volume obtained after using SMC was also investigated. We prepare three samples each of 10 and 100 wt% SMC waste solution. Solutions were neutralized by HCl and the gels separated from the filtered solution. By measuring the radiation intensity then normalizing the readings, we were able to compare the gels and filtered solutions (Fig. 8.5). ^{137}Cs was captured

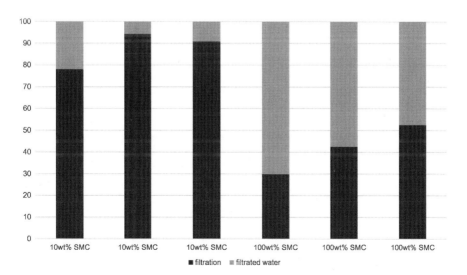

Fig. 8.5 Remediation of ^{137}Cs by neutralizing SMC (10 and 100 wt%)

in the gel after coprecipitation [1, 2]. Hence, each concentration of SMC has the ability to capture Cs effectively. We discovered, however, that the 100 wt% SMC solution had an opposite ability compared with the other solutions. It is because sodium metasilicate can leave its captured ions in water. We need much additional experimentation to determine the optimum conditions for capturing Cs in SMC gels. We are currently trying to neutralize the SMC by alcohol and cleanup ^{137}Cs from waste solutions.

8.4 Conclusions

We have examined the cleaning performance of SMC by comparing the radioactivity of waste solutions before and after washing contaminated granule conglomerates and nonwoven cloth samples. A study of the changes in the chemical characteristics of SMC was also undertaken. ^{137}Cs particulates attached to a nonwoven cloth sample were removed effectively by SMC cleaning. After cleaning, the remaining dissolved aqueous sodium metasilicate from the SMC had a better treatment capacity (after neutralization) than standard aqueous solutions. Thus, the use of SMC should contribute significantly to the decontamination work currently being undertaken in urban and rural areas where decontamination cannot be performed by using water alone. We are currently attempting to measure the chemical structure of SMC by SPring-8. We will be able to report the results of this new study in the near future.

Acknowledgments This work was financially supported by Kyoto University, The Japan Association of National Universities, Japan Association for Chemical Innovation (JACI), ITOCHU Foundation, and the Collaborative Research Program of Institute for Chemical Research, Kyoto University (No. 2013-63). We thank N. Nihei, S. Fujimura, T. Kobayashi, Y. Ono, M. Tosaki, and T. Minami for helpful discussions and experimental preparation. We appreciate the SMC samples (JPAL) provided by Kureha Trading Co. Ltd. The synchrotron radiation experiments were performed at the BL14B2 beamline of the SPring-8 facility with the approval of the Japan Synchrotron Radiation Research Institute (JASRI) (Proposal No. 2015A1662).

References

1. Ueda Y, Tokuda Y, Goto H, Kobayashi T, Ono Y (2014) Removal of radioactive Cs using aqueous sodium metasilicate with reduced volumes of waste solution. ECS Trans 58(19):35
2. Ueda Y, Tokuda Y, Goto H, Kobayashi T, Ono Y (2013a) Removal of radioactive Cs from nonwoven cloth with less waste solution using aqueous sodium metasilicate. J Soc Remed Radioact Contam Environ 1:191

3. Ueda Y, Tokuda Y, Fujimura S, Nihei N, Oka T (2013b) Cesium transfer from granule conglomerate using water containing nano-sized air bubbles. ECS Trans 50(22):1

4. Ueda Y, Tokuda Y, Fujimura S, Nihei N, Oka T (2013c) Removal of radioactive Cs from gravel conglomerate using water containing air Bubbles. Water Sci Technol 67:996

5. Raskin I, Ensley BD (eds) (2000) Phytoremediation of toxic metals: using plants to clean up the environment. Wiley, New York

6. Willey N (ed) (2007) Phytoremediation: methods and reviews. Humana Press, Totowa

7. Agarwal A, Ng WJ, Liu Y (2011) Principle and applications of microbubble and nanobubble technology for water treatment. Chemosphere 84:1175

8. Stride E (2008) The influence of surface adsorption on microbubble dynamics. Philos Trans R Soc A 366:2103

9. Suwabe J (2010) Manufacturing method of a water-based cleaning agent. Japanese Patent (Kokai), 2010-1451

10. Merrill RC, Spencer RW (1950) Gelation of sodium silicate; effect of sulfuric acid, hydrochloric acid, ammonium sulfate, and sodium aluminate. J Phys Colloid Chem 54:806

11. Takahashi M, Kawamura T, Yamamoto Y, Ohnari H, Himuro S, Shakutsui H (2003) Effect of shrinking microbubble on gas hydrate formation. J Phys Chem B 107:2171

Chapter 9
Extractability and Chemical Forms of Radioactive Cesium in Designated Wastes Investigated in an On-Site Test

Yoko Fujikawa, Hiroaki Ozaki, Xiaming Chen, Shogo Taniguchi, Ryouhei Takanami, Aiichiro Fujinaga, Shinji Sakurai, and Paul Lewtas

Abstract In the aftermath of the 2011 accident at Fukushima Daiichi Nuclear Power Plant (F1 hereafter), municipal solid waste (MSW) contaminated with radioactive cesium (rad-Cs hereafter) has been generated in 12 prefectures in Japan. The Japanese Minister of Environment classified MSW that contained rad-Cs in the concentration more than 8,000 Bq/kg as "designated (solid) waste (DSW hereafter), and prescribed the collection, storage and transportation procedures. When MSW containing rad-Cs was incinerated, rad-Cs was concentrated in fly ash, and the ash often fell into the category of DSW. We have investigated a technique that can reduce the volume of the rad-Cs-contaminated fly-ash by extracting rad-Cs with aqueous solvents such as water and oxalic acid and concentrating rad-Cs in a small amount of hexacyanoferrate (or ferrocyanide, designated as Fer hereafter) precipitate. Since DSW could not be transported to the outside laboratory, we have conducted on-site tests at places where DSW were generated to investigate the applicability of the extraction – precipitation technique.

The present report is a summary of our most recent on-site test conducted in 2014. Also presented is the re-evaluation of the results of our past on-site test from the viewpoint of leaching of rad-Cs and heavy metals in the fly ash. An apparent decrease in leaching of rad-Cs from fly ash was observed by incinerating sewage sludge with soil. Fly ash from a melting furnace contained more water-soluble rad-Cs than that from a fluidized-bed incinerator. Some incinerator fly ash appeared to

Y. Fujikawa (✉)
Kyoto University Research Reactor Institute, Asahiro-nishi, Kumatori-cho, Sennan-gun, Osaka 590-0458, Japan
e-mail: fujikawa@rri.kyoto-u.ac.jp

H. Ozaki • X. Chen • S. Taniguchi • R. Takanami • A. Fujinaga
Osaka Sangyo University, 3-1-1 Nakagaito, Daito city, Osaka 574-8530, Japan

S. Sakurai
Osaka Prefecture University, 1-1 Gakuen-cho, Naka-ku, Sakai-shi, Osaka 599-8531, Japan

P. Lewtas
Edith Cowan University, 270 Joondalup Drive, Joondalup, WA 6027, Australia

© The Author(s) 2016
T. Takahashi (ed.), *Radiological Issues for Fukushima's Revitalized Future*, DOI 10.1007/978-4-431-55848-4_9

produce rad-Cs in colloidal form when extracted with oxalic acid, resulting in the lower removal of rad-Cs from the extract by Fer method.

Keywords Ferrocyanide (hexacyanoferrate) • Radioactive cesium • Designated waste • Leaching • Water • Oxalic acid

9.1   Background Information on the Aftermath of the F1 Accident

Four years have passed since the 2011 Eastern Japan Great Earthquake Disaster and the F1 accident that resulted in the first case of widespread contamination of the general environment with rad-Cs in Japan. Millions of people have continued to live in the zone that received rad-Cs fallout higher than the level of background global fallout observed in the middle latitudes of the northern hemisphere [1, 2]. Public and semi-public services indispensable for daily life, e.g. supply of electricity and potable water, sewage treatment, garbage collection, public transportation, etc., had been reconstructed relatively quickly after the 2011 disaster (e.g. [3, 4]). Among these services in the affected areas, potable and sewage water treatment and garbage collection have produced drinking water treatment sludge, sewage sludge, and incinerator ashes containing elevated concentration of rad-Cs. Both disposal and reuse have become difficult for such wastes, resulting in sludge and ashes piling up in many treatment facilities.

The government of Japan has proposed to construct landfills with special leachate collection and treatment systems along with a radiation monitoring system for the disposal of wastes with rad-Cs concentration higher than 8,000 Bq/kg (so-called DSW). But so far the plan has stalled due to public criticism. We have summarized the situation in our previous reports [5, 6], and the situation has remained unchanged concerning the final disposal of DSW.

9.2   Strategy of Volume Reduction and Stabilization of Municipal Solid Waste and Designated Waste in Japan

Under the circumstances delineated above, the volume reduction of various wastes contaminated with rad-Cs is of paramount importance. The thermal treatment, with emphasis on incineration, has been the topmost strategy for stabilization and volume reduction of municipal and industrial wastes in Japan [7]. Also, thermal treatment has been a recommended option for volume reduction of combustible radioactive wastes for a long time [8]. Consequently, incineration has been planned and conducted on vegetation and paddy straw removed in the course of the cleanup work and sewage sludge contaminated with rad-Cs from the area with high rad-Cs fallout, resulting in generation of bottom ash and fly ash possessing diverse leaching characteristics of rad-Cs and non-radioactive hazardous components.

In Japan, various pretreatment procedures have been studied and administered for incinerated MSW before burying the waste in landfills [9, 10]. The purpose of the pretreatment had been to accelerate the waste stabilization, namely to prevent the leaching of toxic elements, hazardous organic substances and dissolved and/or suspended organic matters related with BOD and COD values, from the waste. Among such procedures was the washing of ashes with acid or water (e.g. [11, 12]). The washing effectively removed the major elements from MSW incineration ashes and also reduced the leachability of toxic elements from the washed residues. In a full-scale acid extraction process, washed ash became a stabilized cake to be buried in a landfill, solid salt was recycled to soda production plants, condensed water was recycled in the plant as cooling water and no wastewater was discharged from the plant [11]. More recently, a group of researchers and engineers proposed the "wash out waste" (WOW) system in which solid waste was inactivated by washing before landfilling, resulting in rapid stabilization of the buried waste and early closure of the landfill [13]. The WOW system can reduce the possibility of long-term contamination of the environment by landfill leachate, and may also help gain public support for the siting of landfills.

After the F1 accident and the concomitant pollution of MSW with rad-Cs, waste-washing technology has drawn the attention of researchers and private companies again. The nature of Cs in incinerated (or in some case smelted) ashes and the decontamination technology to remove rad-Cs from the wastes or waste extracts have been investigated [5, 6, 14–16]. Parajuli et al. [14] pointed out that the percentage of water-soluble rad-Cs is much smaller in sewage ash compared to that in wood and garbage ash, probably because of clay minerals of pedogenic origin present in the sewage sludge. Saffarzadeh et al. [15] studied the behaviour of Cs in bottom ash using stable Cs salts in a pilot incinerator, and reported that Cs in bottom ash could be most commonly found in the silicate glass matrix. Kozai et al. [16] studied the mineralogical components of sewage sludge ashes and reported that the majority of rad-Cs was in the HCl-soluble phase, and was probably contained in iron oxides.

The fly ash washing technique to eliminate rad-Cs has been also studied by some private companies, and the initial and running cost of the technology has been estimated [17]. In the processes developed by the companies, the use of Fer compounds to remove rad-Cs from the washing solution was a prerequisite. A report by the Institute of National Environmental Studies [18] defined the category of rad-Cs containing wastes for which washing processes can be applied and the performance requirements as well as radiation safety standard of the washing system.

9.3 Principle of Ferrocyanide Coprecipitation for Cs Removal

The principle of the Fer method for removal of Cs is reviewed in literatures [19–21], as well as in our previous report [5, 6]. In brief, the method utilizes the extremely high affinity of the Cs cation to insoluble Fer compounds. The insoluble Fer compounds are prepared by reacting soluble Fer salts (K, Na or H compounds) with metal (Fe, Cu, Zn, Ni, Cd, Mn, etc.) ions in solution.

There are three different ways to use Fer in order to remove Cs: (1) addition of soluble Fer salts and metal elements to waste solution (in situ formation of Fer solid), (2) addition of freshly prepared insoluble Fer metal complex slurry to waste solution and (3) use of Fer-metal adsorbents in solution. The distribution ratio of Cs to insoluble Fer compounds is high in the order of method (1) > (2) > (3). Throughout our experiment, we have used method (1) (in situ method) as much as possible, and have used method (2) (addition of the fresh Fer solid) when method (1) could not be used, e.g. when Zn concentration in the solution is very high and Zn-Fer was going to be the prevalent Fer solid to be formed.

9.4 The Waste Volume Reduction When Ferrocyanide Coprecipitation Technique Was Used

We have been working on a technique that extracts rad-Cs from DSW (especially rad-Cs contaminated fly ash) with water or hot oxalic acid, and selectively removes rad-Cs from the extract as a small amount of precipitate utilizing the Fer coprecipitation technique. The focus of our research has been to achieve the efficient removal of rad-Cs from the waste extracts having a complex composition. So far, we have found that more than 95 % of rad-Cs could be removed from the extracts of MSW under an optimized combination of pH, Fer concentration and type of metal-Fer complex formed in the extract. A high Zn concentration in the extract suppressed the removal of Cs by the Fer precipitation method [5, 6].

The waste volume reduction d [%] and the ratio of rad-Cs concentration in the precipitate to that of the original waste Q_{Cs}/C_0 are estimated by the following equation:

$$d = 100 \left(1 - \frac{pV}{M} \right) \tag{9.1}$$

$$\frac{Q_{Cs}}{C_0} = \frac{100}{100 - d} rE \tag{9.2}$$

where M: weight [kg] of MSW extracted by V [L] of the solvent, p [kg/L]: weight of Fer precipitate formed per unit volume of the extract, Q_{Cs} [Bq/kg]: rad-Cs concentration [Bq/kg] in Fer precipitate, C_0 [Bq/kg]: rad-Cs concentration in original MSW, r: ratio of rad-Cs removed from the extract of MSW by Fer technique, and E: ratio of rad-Cs extracted from MSW with solvent.

We estimated the volume reduction d (%) experimentally by a cold run using fly ash of sewage sludge obtained from the area unaffected by rad-Cs from the F1 accident [22]. The procedures are as follows: M [kg] of fly ash samples was extracted with V [L] of oxalic acid or distilled water, 100 μg/L of Cs-133 (stable isotope of Cs) as CsCl salt was added to the extract as a spike, and Fer coprecipitation was conducted to remove Cs from the extract. After the coprecipitation, the precipitate was separated from the liquid by centrifugation (6,400 g, 4 °C, 15 min), freeze dried and weighed to estimate the value of p [kg/L]. The Cs concentration in the supernatant after Fer coprecipitation was determined by ICP-MS to estimate the Cs removal (r in Eq. 9.2). The d (%) values for actual DSW used in our on-site test were not obtained due to the shortage of time during the test.

The results obtained are summarized in Table 9.1. When concentration of Fer added for the removal of Cs was 0.1 mM, the theoretical d (%) value is 99.3 % for oxalic acid extract and 99.99 % for water extract, under the assumptions that the solid-to-liquid ratio of the extraction of waste was 1 g/2.5 mL, that the extract was diluted by a factor of 100 when the extracting reagent was 1 M oxalic acid (factor of 2 when the reagent was water) before subjecting it to Fer coprecipitation and that the Fer solid is anhydrous. As shown in the table, the observed d (%) values were generally higher than 99.9 % for the water extract and were close to the theoretical value. On the other hand, d (%) values were low for oxalic acid extract especially when Fer coprecipitation was conducted at circumneutral or alkaline pH. The original oxalic acid extract had low pH (c.a. 1), and as we added NaOH to increase the pH of the solution, the precipitates of metal hydroxides were formed. It was also noted that the oxalic acid extract of fly ash KO had a lower volume reduction compared to that of fly ash O under the same pH values. This is ascribed to the high amount of Fe contained in the extract of fly ash KO [6], starting to precipitate around pH 3. In summary, for the oxalic acid extract, higher waste volume reduction is achieved by conducting the Fer coprecipitation at pH value as low as possible so that the formation of metal hydroxide precipitate can be suppressed.

Overall, the results in Table 9.1 indicate that Fer coprecipitation leads to significant reduction in the amount of the waste (more than 1/1,000 for waste extracted with water, or nearly 1/100 for waste extracted with oxalic acid), under appropriate operating conditions. The waste residue after the extraction contains a low concentration of rad-Cs, and, since the extractable rad-Cs was removed from the residue, the long-term contamination of the environment caused by leaching of rad-Cs from the buried waste residue is expected to be negligible. Combined with proper management (e.g. storage) of the small volume of rad-Cs concentrated waste, the siting of the final disposal site of DSW is expected to become much easier.

Table 9.1 Volume reduction of the waste by extraction-precipitation method (results of cold run)

Extracting reagent	Fer [mM][a]	pH	Fly ash O			Fly ash KO		
			d %[b]	r	$(Q_{Cs}/C_0)/E$	d %[b]	r	$(Q_{Cs}/C_0)/E$
Distilled water[c] 1 g/2.5 mL solid to liquid ratio	0.1	3	n.a.	0.959	n.a.	99.9<	0.952	667<
		4	99.933	0.946	1,401	99.916	0.990	827
		5	99.9<	0.958	671<	99.935	0.959	1,031
		6	99.9<	0.933	653<	99.985	0.995	4,642
		7	99.961	n.a.	n.a.	99.963	0.992	1,851
		8	99.848	0.988	454	99.965	1.000	1,999
		10	99.911	0.985	63	99.713	0.993	242
	0.5	5	99.9<	0.974	682<	99.970	0.989	2,305
	1.0	5	99.9<	0.942	659<	99.961	0.999	1,793
1 M oxalic acid[d] 1 g/2.5 mL solid to liquid ratio	0.1	3	98.66	0.999	74	82.33	0.998	6
		4	94.41	0.996	18	77.45	0.998	4
		5	90.32	0.969	10	74.04	0.957	4
		6	86.58	0.978	7	78.65	0.937	4
		7	83.69	0.960	6	70.65	0.963	3
		8	77.82	0.926	4	63.43	0.962	3
		10	76.79	0.924	4	76.11	0.919	4
	0.5	5	97.68	0.998	43	69.35	0.985	3
	1.0	5	88.07	0.988	8	69.74	0.928	3

[a]When 0.1 mM of Fer was added to the extract resulting in $Fe_4[Fe(CN)_6]_3$ or $Ni_2[Fe(CN)_6]$ precipitate, the weight of the precipitate is 29 or 34 mg/L respectively (excluding the weight of the hydrated water)

[b]Theoretically, volume reduction d (%) is 99.3 % for oxalic acid extract and 99.99 % for distilled water extract based on the weight of the anhydrous Fer solid

[c]The extract was diluted by a factor of 2 before subjecting it to the Fer coprecipitation

[d]The extract was diluted by a factor of 100 before subjecting it to the Fer coprecipitation

9.5 Objectives of the Present Study

The extraction-coprecipitation method described above can be an ultimate solution for volume reduction of DSW. The waste volume reduction d (%) in Eq. 9.1 is an important factor in assessing the effectiveness of the method, but also important is the relative concentration of the rad-Cs in the precipitate compared to that in the original waste. Higher values of Q_{Cs}/C_0 imply that rad-Cs is concentrated more efficiently in the precipitate compared to the waste residues after the extraction, thereby defining the overall effectiveness of the method.

The starting point of our study has been to identify the factors that are likely to govern the rad-Cs removal (i.e. r in Eq. 9.2) from DSW extracts by Fer coprecipitation technique, and to optimize coprecipitation conditions for Cs removal. Based on this perspective, we evaluate the results of our most recent (fiscal year 2014) on-site test using DSW in the present report.

Leachability of rad-Cs in the DSW (i.e. E in Eq. 9.2) is another important factor that governs the effectiveness of the extraction-coprecipitation method. We therefore evaluated the results of our 2013 and 2014 on-site tests to examine the leaching of rad-Cs from sewage sludge ashes. Also important are the leaching characteristics of heavy metals. High leaching of Fe and Ni from waste is convenient in our study since these metals can be effectively used for the formation of insoluble Fer precipitate. On the other hand, high leaching of Zn is inconvenient because Zn forms insoluble Fer precipitate when soluble Fer salt is added to the solution, but the removal efficiency of Cs by Zn-Fer is low [6]. In this context, in the present report, we also examine the leaching of heavy metals from the fly ash samples.

9.6 Experimental

9.6.1 Reagents

All reagents used in our study were of analytical grade unless specified otherwise.

9.6.2 Procedures of Cold Run

We used MSW fly ash samples unaffected by the F1 accident to clarify the effect of extracting procedures on the extraction efficiency of various elements (including non-radioactive cesium) from the sample. In particular, we compared the use of 0.1 and 0.5 M hot oxalic acid extraction procedures on fly ash from a melting furnace treating sewage sludge from Western Japan. Since the sample was not contaminated with rad-Cs from the F1 accident, the extraction could be conducted in the university laboratory, and the concentration of various elements in the extracts could be determined using ICP-MS (model 7700 series, Agilent Technologies).

The three different extraction procedures were tested on the fly ash sample from the melting furnace. (1) One procedure was vacuum extraction (or continuous extraction), in which a 5 g fly ash sample placed on an i.d. 47 mm membrane filter holder (fitted with pore size 0.45 μm PTFE filter) was continuously eluted with hot oxalic acid (50 mL solution/40 min) supplied via a peristaltic pump for total 120 min. The filter holder was connected to a vacuum flask, and vacuum was applied periodically so that the extracting reagent (0.1 or 0.5 M hot oxalic acid) was passed through the filter while the fly ash sample was always soaked in the oxalic acid. The percolate was collected every 40 min, and was analyzed by ICP-MS. By using the vacuum extraction method, re-adsorption of once extracted components to the waste material, as well as temporary oversaturation of extracted metals in the solution phase, could be minimized. This procedure was also practicable in the on-site test. (2) Conventional batch extraction (2.5 g of fly ash in 75 mL of 0.1 or 0.5 M hot oxalic acid), with 200 rpm of shaking in a 90 °C hot bath, was conducted. (3) Also tested was a batch exchange extraction method, an intermediate method between continuous extraction and batch extraction. With this method, 2.5 g of fly ash was extracted with 25 mL of 0.1 M or 0.5 M hot oxalic acid with continuous shaking (200 rpm) in a hot bath for 1 h, centrifuged for solid–liquid separation (6,400 × g, 4 °C, 15 min), and the supernatant was removed and saved for the subsequent analysis by ICP-MS. Then 25 mL of 0.1 or 0.5 M hot oxalic acid was added to the solid waste again and the same extraction procedure repeated twice. The batch exchange method turned out to be laborious and was not practicable in the on-site test compared to the other two methods.

With hot oxalic acid extraction, a precipitate was sometimes formed in the extract after it was cooled to room temperature. In such cases, the extract was centrifuged, the precipitate was dissolved in dilute HNO_3 and the supernatant and the dissolved precipitate were analyzed separately by ICP-MS.

To determine the total element concentration in the fly ash, 100 mg of the fly ash sample, pulverized to pass a 150 mesh sieve, was digested by mineral acids in a pressurized Teflon vessel placed in a microwave oven model MARS5000 (CEM Co., US). The acids used were 5 mL of concentrated 68 % HNO_3, 1 mL of 70 % $HClO_4$ and 4 mL of 38 % HF. All the acids used were AA-100 grade from Tama Chemicals Corporation (Japan). The sample was acid decomposed under 70 psi pressure for 10 min. after 20 min. of ramp time. After the decomposition, 6 mL of 4.1 w/v percent boric acid was added to the digested sample to dissolve the fluoride precipitated with Ca, etc. as well as to neutralize HF left in the solution. With this procedure, a clear solution with no visible suspended particles was obtained. The solution was diluted and analyzed by ICP-MS. The results obtained were used to calculate the percentages of the elements extracted by 0.1 or 0.5 M oxalic acid from the fly ash per total amount of the elements in the original fly ash.

9.6.3 Determination of Radioactive Cs and Non-radioactive Metals in the On-Site Test

On-site analysis of rad-Cs was conducted using a portable Ge detector from Princeton Gamma Technology (P-type, relative efficiency 20 %), and a Ge detector from ORTEC (model Trans-Spec, p-type, relative efficiency 40 %), in 2013 and 2014, respectively. Ge detectors were pre-calibrated for absolute full-energy peak efficiency for the solid samples packed in commercially available U-8 vessels (i.d. 50 mm, height 68 mm) as well as for a 30 mL volume of aqueous solution in a 50 mL polypropylene bottle, and were used to quantitate rad-Cs in waste extracts and waste materials. A NaI detector (model AN-OSP-NAI, Hitachi-Aloka-Medical, pre-calibrated for determination of rad-Cs) was also used to quantitate Cs-134 and Cs-137 in waste materials.

Concentrations of metals in the extract obtained at the on-site test were determined using a portable voltammetry instrument (model PDV6000plus from Modern Water, UK) coupled with an autosampler. In 2014, Fe and Zn were determined. Determination of Fe was conducted using a glassy carbon working electrode, acetate buffer as a supporting electrolyte, deposition potential $-1,600$ mV and stripping step potential range of $-1,100$ to 50 mV. Determination of Zn was conducted at a mercury-film-coated glassy carbon working electrode, acetate buffer as a supporting electrolyte, deposition potential $-1,300$ mV and stripping step potential range of $-1,200$ to 50 mV.

9.6.4 Extraction Procedures Used in the On-Site Test

The rad-Cs-contaminated fly ash samples from sewage treatment in the area affected by the F1 accident were extracted and analyzed for rad-Cs in the plant where they were produced (the on-site test). The information for the waste materials used for the on-site test is summarized in Table 9.2, together with extracting reagents tested on the samples. We used a distilled-water extraction [23] and a hot oxalic acid (0.1 M) extraction [24] that had previously been applied to the removal of rad-Cs from contaminated MSW. Also tested were 5 M HCl extraction and 0.5 M hot oxalic acid extraction in an attempt to achieve higher leaching of rad-Cs for some samples.

The extraction of rad-Cs containing waste materials was conducted using the vacuum extraction (continuous extraction) procedure in the on-site test conducted in 2014. This procedure was used because the results of three different extraction methods for uncontaminated waste showed that the batch extraction may be affected by loss of the extracted elements due to re-adsorption and/or precipitation processes as will be discussed in the Sect. 9.7. The extraction in 2013 was conducted by batch extraction.

9.6.5 Ferrocyanide Coprecipitation Procedures Used at the On-Site Test

The treatment of the extract by Fer precipitation technique was conducted as follows: The pH of the waste extract containing rad-Cs was adjusted to the desired values (shown later in Sect. 9.7) using NaOH and HCl. Potassium ferrocyanide ($K_4[Fe(CN)_6]$, K-Fer hereafter) was used as a soluble Fer salt source. Since metal hydroxides (ineffective for Cs removal) rather than metal-Fer complexes could be formed in the solutions when pH values are in neutral to alkaline regions, a slightly acidic pH was used for Fer precipitation. After the addition of K-Fer, Ni-Fer or Fe(III)-Fer precipitate (prepared in a separate bottle) to the extract, the sample was

Table 9.2 Designated waste samples subjected to the on-site test

Sample code	Type of waste material	Period of waste generation	Period of time when on-site test was conducted	rad-Cs concentration [kBq/kg][a]		Extracting reagent tested
				Cs-134	Cs-137	
IF1	Incinerator fly ash[b]	October, 2013	October, 2013	4.0	8.5	Distilled water[c] and 0.1 M hot oxalic acid
IF2	Incinerator fly ash[d]	October, 2014	October, 2014	2.7	8.1	Distilled water, 0.1 M hot oxalic acid, 0.5 M hot oxalic acid, and 5 M HCl
IB	Incinerator bottom ash[e]	October, 2013	October, 2013	0.6	1.2	Distilled water[c] and 0.1 M hot oxalic acid
MF1	Fly ash from melting furnace[f]	May, 2013	October, 2014	14.5	45.4	Distilled water, distilled water (pH3), 0.1 M hot oxalic acid and 0.5 M hot oxalic acid
MF2		October 2013	October, 2013	32.0	67.4	Distilled water and 0.1 M hot oxalic acid
MF3		September, 2014	October, 2014	6.2	19.5	Distilled water (pH3), 0.1 M hot oxalic acid and 0.5 M hot oxalic acid

[a]Radioactivity at the time of the experiment
[b]Sewage sludge and the cover soil were incinerated together in a fluidized bed incinerator at 850 °C. Volume reduction by incineration was approximately 67 %
[c]The solution pH was adjusted to 3 during the extraction by titrating with HCl
[d]Only sewage sludge was incinerated. Volume reduction of the sludge by incineration was approximately 95 %
[e]Incinerated soil was discharged as a bottom ash of the fluidized bed incinerator
[f]Sewage sludge was smelted at 1,200 °C. Volume reduction of the sludge by the smelting was approximately 97 %

centrifuged at $2,600 \times g$ at 4 °C for 20 min. or stood still until the precipitate was settled. When more than a few liters of the extract were treated, 30 mg/L of cationic polymeric flocculant (polymethacrylic ester, C-303 or C-512 from MT AquaPolymer Inc., Japan) was added to facilitate the settling of fine precipitates.

9.7 Results and Discussion

9.7.1 Extraction of Metal Elements from Fly Ash Sample in the Cold Run

Extractions of stable (non-radioactive) Ca, Fe, Zn, Ni, As and Cs from fly ash (collected from a melting furnace treating sewage sludge from Western Japan) with 0.1 and 0.5 M hot oxalic acid are compared in Fig. 9.1. Clearly, the percentages of metal extraction differed considerably depending on the concentration of oxalic acid as well as on the extraction procedures (continuous, batch and batch exchange). With the batch extraction procedure, we observed a high extraction percentage at an earlier time (40 min), followed by a decrease in the extraction percentage. This is probably explained by re-adsorption or re-precipitation (or co-precipitation) of once extracted elements. In contrast, the continuous extraction showed a gradual increase in extraction percentages with time. The extraction percentages by continuous extraction at 120 min of extraction time were similar to those obtained by batch extraction at 40 min for non-radioactive Cs and Fe. The batch exchange procedure method showed a gradual increase in the extraction percentages of the elements with time and was more similar to the continuous extraction than simple batch extraction. For both batch and batch exchange procedures, there could have been temporary oversaturation of the extracted elements in the solution after 40 min of extraction with hot oxalic acid. In the case of the batch exchange method, the oversaturated elements, if any, should have been precipitated during the centrifugation to exchange the supernatant. In simple batch extraction, unlike batch exchange extraction, supernatant was not exchanged, and the aliquot of the supernatant collected for analysis should contain the oversaturated element (if any). In any case, the batch exchange procedure was laborious compared to the continuous extraction and simple batch methods. Also, the centrifugation we conducted for solid-liquid separation in the batch exchange procedure every 40 min caused a temporary decrease in the extraction temperature, making it difficult to interpret the obtained results.

The oxalic acid concentration of the extract also affected the extraction efficiency of elements (in Fig. 9.1, only the results of Ca, Fe, Zn, Ni, As and Cs are shown). Briefly, in the case of continuous extraction, 0.5 M hot oxalic acid was more effective than 0.1 M hot oxalic acid in extracting non-radioactive Cs, Fe, Mg, Al, Mn, As, Se, Sr and Ba. On the other hand, for Zn, Ca, Co, Ni and Cu, 0.1 M hot oxalic acid was more effective than 0.5 M hot oxalic acid (in the case of continuous

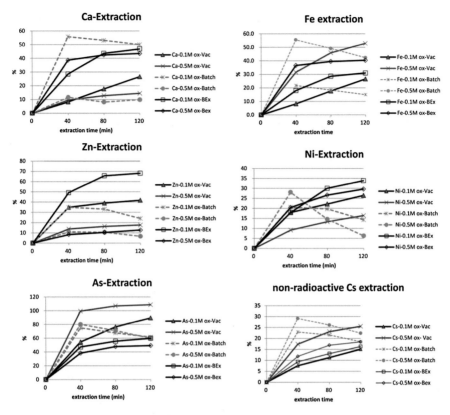

Fig. 9.1 Comparison of extraction rate of various elements with continuous (vacuum extraction), batch extraction, and batch exchange extraction methods for fly ash from a melting furnace. Here, 0.1 M ox − Vac: continuous extraction with 0.1 M hot oxalic acid, 0.5 M ox − Vac: continuous extraction with 0.5 M hot oxalic acid, 0.1 M ox-Batch: batch extraction with 0.1 M hot oxalic acid, 0.5 M ox-Batch: batch extraction with 0.5 M hot oxalic acid, 0.1 M ox − Bex: batch exchange extraction with 0.1 M hot oxalic acid, and 0.5 M ox − Bex: batch exchange extraction with 0.5 M hot oxalic acid

extraction). Oxalic acid dissolves Fe and some other metals preferentially through the ligand-promoted dissolution process [25], and the lower pH of 0.5 M oxalic acid compared to 0.1 M oxalic acid enhances the dissolution of elements in general. At the same time, with 0.5 M oxalic acid, metal oxalates having low solubility can be more easily formed than with 0.1 M oxalic acid. We consider that the concentrations of Zn, Ca, Co, Ni and Cu in the 0.5 M oxalic acid extract were solubility controlled by the metal oxalates. Although the solubility values of metal oxalates freshly formed at high temperature (90 °C) may be hard to obtain, solubility determined for pure metal oxalates can be used as reference values. According to the literature [26], oxalates of Zn, Co, Ni and Cu have very low solubility at 20 °C. This information supports our inference that Zn, Co, Ni and Cu concentrations were solubility controlled. Since the concentration of Ca in the extract is high (between 67 and 600 mg/L in 0.5 M oxalic acid extract), it is no wonder that the Ca concentration was controlled by the solubility of calcium oxalate.

Based on the results above, we applied the continuous extraction method at the on-site test in 2014, in order to obtain more scientific estimates of the leaching of rad-Cs from the contaminated fly ash. We noted, however, that the behavior of non-radioactive Cs during the extraction was relatively simple, i.e. Cs was not precipitated with oxalic acid and therefore its extraction was higher in 0.5 M oxalic acid than in 0.1 M oxalic acid, and the highest extraction could be achieved with 40 min of batch extraction. From a practical point of view, by using 0.5 M hot oxalic acid in simple batch extraction mode, the majority of Cs that can be extracted with the reagent by the continuous method will be released from the solid phase, although there is a chance that the behavior of non-radioactive Cs we have investigated in this section and that of rad-Cs in fly ash samples are different, depending on the original chemical form of Cs.

9.7.2 Extraction of rad-Cs from Contaminated Fly Ash Samples

The results are summarized in Table 9.3. The results indicate rad-Cs leaching from melting furnace fly ash (MF1, MF2 and MF3) is high. At least 40 % was extracted with distilled water, and the percentage of the rad-Cs extraction was close to 100 % when 0.5 M hot oxalic acid was used. Notably, the percentages of rad-Cs extraction were not much affected by the pH values of the extract. For example, distilled water extracts of MF3 had pH values around 7 (refer to Table 9.3), and rad-Cs extraction was higher than 60 %. The distilled water extracts of MF2 had lower pH values (around 1) compared to MF3, and the rad-Cs extraction was only 40 %, lower than the distilled water extracts of MF3 with pH value 7. The lower extraction percentages for MF2 are partly attributable to the lower solid-to-liquid ratio (1 g/2.5 mL) of extraction compared to 1 g/75 mL for MF3, implying that a thorough washing with larger volume of liquid leads to higher leaching.

As discussed in the preceding section, stable Cs extracted by distilled water from fly ash of the melting furnace treating sewage sludge from Western Japan was only 5 % of the total Cs contained in the ash. Interestingly, rad-Cs in the fly ash MF1, MF2 and MF3 from the similar melting furnace was more water soluble compared to the stable cesium in the other fly ash of similar origin.

The extractability of rad-Cs was lower for incinerator fly ash (IF1 and IF2) compared to fly ash from the melting furnace (MF1, MF2 and MF3). The ashes collected in different periods, IF1 and IF2, also exhibited quite different leaching of rad-Cs, Fe and Zn. When extracted with distilled water, rad-Cs extraction percentages were only ca. 1 % for IF1, whereas it was 10 % or higher for IF2. The IF2 ash also had higher amount of extractable Fe and Zn compared to IF1. The IF1 ash was obtained when sewage sludge was incinerated together with soil, while IF2

Table 9.3 Results of rad-Cs extraction by various extracting reagents

Sample code	Extracting reagent	Extraction time	pH	Fe (mg/g-soil)	Zn(mg/g-soil)	Percentage of Cs extracted	Extracting procedures and solid/solution ratio
IF1	Distilled water (titrated to pH3)	1 h	3.0 (8.9 before pH adjustment)	–	–	2.1 %	Batch, 1 g/2.5 mL
		2 h	3.9	–	–	0.4 %	
		3 h	3.5	–	–	0.7 %	
	0.1 M hot oxalic acid	1 h	4.2	0.08	0	1.3 %	
	0.5 M hot oxalic acid	1 h	1.0	0.63	0.002	11 %	
		2 h	–	–	–	11 %	
		3 h	–	–	–	10 %	
	5 M HCl	1 h	0	–	0.023	24 %	
		2 h	–	–	–	20 %	
		3 h	–	–	–	25 %	
IF2	Distilled water	40 min	5.8	0.0	–	14 %	Continuous, 1 g/30 mL
		80 min	6.9	0.0	–	14 %	
	Distilled water (pH3)	40 min	5.5	0.0	0.2	12 %	
		80 min	6.7	0.0	–	12 %	
		120 min	6.8	–	–	19 %	
	0.5 M hot oxalic acid	40 min	1.2	43.4	13.4	39 %	
		80 min	0.7	–	–	61 %	
		120 min	0.7	–	–	66 %	
IB	Distilled water	1 h	3.0 (5.1 before pH adjustment)			0 %	Batch, 1 g/2.5 mL
		2 h	3.1			0 %	
		3 h	3.3			6 %	
	0.1 M hot oxalic acid	1 h	1.5	0.01	0.0	2.9 %	

	Extractant	Time					Method
MF1	Distilled water	40 min	Not analyzed	–	–	54 %	Continuous, 1 g/25 mL
		80 min	Not analyzed	–	–	58 %	
	Distilled water (pH3)	40 min	6.0	0.0	0.0	66 %	
		80 min	6.1	0.0	0.0	70 %	
	0.1 M hot oxalic acid	40 min	Not analyzed	0.5	–	85 %	
		80 min	Not analyzed	5.0	–	100 %	
	0.5 M hot oxalic acid	40 min	0.7	1.1	2.6	100 %	
		80 min	0.7	1.5	3.1	100 %	
MF2	Distilled water	1 h	1.7	–	–	41 %	Batch, 1 g/2.5 mL
		1.5 h	Not analyzed	–	–	46 %	
		3 h	2.0	–	–	42 %	
	0.1 M hot oxalic acid	1 h	1.7	2.3	2.2	59 % and 65 % (n = 2)	Batch, 1 g/5.0 mL
MF3	Distilled water	40 min	7.6	–	–	64 %	Continuous, 1 g/25 mL
		80 min	7.1	–	–	69 %	
	Distilled water (pH3)	40 min	7.2	0.0	0.0	59 %	
		80 min	6.0	–	0.0	61 %	
		120 min	6.0	–	0.0	61 %	
	0.1 M hot oxalic acid	40 min	1.6	0.46	–	67 %	
		80 min	1.4	1.88	–	77 %	
	0.5 M hot oxalic acid	40 min	0.8	–	32[a]	73 %	
		80 min	1.8	–	42[a]	81 %	
		120 min	0.8	–	49[a]	88 %	

[a]Reevaluation of the data is under way

ash was obtained when only sewage sludge was incinerated. The other researchers [9] pointed out that the incineration with soil is likely to inhibit the leaching of rad-Cs from the fly ash, and the results we obtained are in accordance with their observation. As for IF1, we also used 5 M HCl extraction in an effort to facilitate the leaching of rad-Cs, but the percentage of rad-Cs extracted was only 20 % with this method. Kozai et al. [11] reported that leaching of rad-Cs from sewage ash can be enhanced by pulverizing the ash and using hot HCl extraction. Out of concern for safety, we could not conduct the pulverizing of already fine-grained, powder-like incinerator fly ash nor use hot HCl during our on-site test. In contrast, more than 60 % of rad-Cs in IF2 was extracted with 0.5 M hot oxalic acid (the percentage was 10 % for IF1). We cannot rule out the possibility that the continuous extracting procedure we used for IF2 (IF1 was extracted by batch method) was more effective in leaching rad-Cs from the sample. However, as we discussed in the section above, the leaching behavior of Cs is relatively simple because of the high solubility of Cs in water, and it is unlikely that the difference in extraction procedure (continuous or batch) led to a drastic difference in the extraction rate with the same hot oxalic acid. We therefore conclude that the incinerating of sewage sludge and soil together in the fluidized bed incinerator led to the drastic reduction of rad-Cs leaching from fly ash.

The bottom ash IB1, which in fact is the soil incinerated in the fluidized bed incinerator, exhibited a very low leaching of rad-Cs (0 % with distilled water and 3 % with 0.1 M hot oxalic acid). Considering the reduction in leaching of rad-Cs from fly ash, it is obvious that rad-Cs leaching was inhibited in the presence of soil. It should be noted that the soil itself did not contain much rad-Cs before it was incinerated. The rad-Cs contained in the bottom ash originated from contaminated sewage sludge.

9.7.3   Ferrocyanide Coprecipitation in On-Site Test

Table 9.4 shows results of Fer coprecipitation conducted on extracts of rad-Cs contaminated fly ash. The extract of molten fly ash showed c.a. 93 % removal of rad-Cs when Fe(III)-ferrocyanide or Ni(II) –ferrocyanide was added and the supernatant was analyzed. Removal of rad-Cs, when we tried the in situ formation of insoluble Fer solids by adding Fe(III) salt and K-Fer salt, was only 80 % because the very fine precipitate formed could not be completely removed, even though we added a coagulant. Optimal solid-liquid separation for the in situ Fer solid formation method needs to be pursued further.

As for the extracts of incinerator ash IF2, only one oxalic acid extract sample contained an easily detectable amount of rad-Cs (235 Bq/L). We therefore focused on the removal of rad-Cs from this sample. After some trial and error, we conducted coprecipitation with Ni(II)-ferrocyanide at pH7. The resultant removal of rad-Cs was only 59 %.

Table 9.4 Removal of rad-Cs from the solution under different pH and ferrocyanide concentrations (FY 2014 on-site test results)

Sample type	Cs-137 Bq/L in raw water	pH	Ferrocyanide concentration	Metal added	Removal %
0.3 M oxalic acid extract of fly ash from a smelting furnace	800	2	0.6 mM iron ferrocyanide	(Fe(III)-ferocyanide)	93.8 %[a]
Water extract of fly ash from a smelting furnace	842	Between 3 and 4	0.1 mM potassium ferrocyanide	0.4 mM Fe(III)	80.9 %
Water extract of fly ash from a smelting furnace	842	7	0.1 mM nickel ferrocyanide	(Ni(II)-ferrocyanide)	92.50 %
0.5 M oxalic acid extract of fly ash from an incinerator	33	Between 2 and 3	0.1 mM iron ferrocyanide	(Fe(III)-ferrocyanide)	100 %
					Cs-137 n.d. after treated
Concentrated 0.5 M oxalic acid extract of fly ash from an incinerator	235	7	0.1 mM nickel ferrocyanide	(Ni(II)-ferrocyanide)	58.8 %[b]
Water extract of fly ash from an incinerator	15	3.9	0.1 mM iron ferrocyanide	(Fe(III)-ferrocyanide)	100 %
					Cs-137 n.d. after treated

[a]Before centrifugation. Removal was improved to 95.8 % after centrifugation
[b]The result of ultrafiltration suggested that about 34 % of radioactive cesium was in the particulate form

To investigate the cause of this low removal of rad-Cs, in the 2014 on-site test, the 0.5 M oxalic acid extracts of fly ashes from the incinerator and melting furnace (both pre-filtered by a pore size 0.45 μm membrane filter) were ultra-filtered using molecular weight (MW) 3,000 Sartorius centrifugal ultrafiltration kit. The extract of fly ash from the melting furnace left no residue on the ultrafilter, indicating that there are probably no colloidal particles in the extract of the melting furnace. On the other hand, the extract of the incinerator ash was retained on the ultra-filter, and by analyzing the rad-Cs content of the retained liquid, c.a. 30 % of rad-Cs in the extract was estimated to be colloidal (MW >3,000).

These results explain why rad-Cs removal from the 0.5 M oxalic acid extract of the incinerator fly ash by Fer coprecipitation was low (58 %). Only rad-Cs cations can be removed by the Fer precipitation technique, and if the rad-Cs in colloidal form in the incinerator fly ash extract had been in cationic form, the removal of rad-Cs would have increased to c.a. 90 %, i.e. the removal that should be attained with the Fer method under normal circumstances. As for the extract of incinerator fly ash, we also observed the low removal of rad-Cs in the on-site test in 2013 and attributed the results to the lack of time to optimize the coprecipitation conditions for the sample [6]. However, based on the recent findings, the low removal of rad-Cs in the 2013 test was also likely to have been the effect of rad-Cs being present in colloidal form. Although the conversion of the colloidal rad-Cs to rad-Cs cations may not be easy, by selecting and using an appropriate coagulant, colloidal rad-Cs can be removed by coagulation and precipitation. The origin of the colloidal rad-Cs in the extract needs to be investigated in the future research.

9.8 Conclusion

Extraction of fly ashes generated from sewage treatment plants in the area affected by the F1 accident was investigated, followed by the selective removal of Cs using Fer precipitation techniques. The extraction of Rad-Cs was lower for incinerator fly ash compared to fly ash from the melting furnace. When sewage sludge was incinerated with soil, leaching of rad-Cs from the fly ash was significantly reduced. It was also noted that the leaching of rad-Cs from fly ash of the melting furnace was probably higher than that of stable Cs. Colloidal (non-ionic) rad-Cs, found in incinerator fly ash extract, reduced the Cs removal by Fer precipitation technique. The use of an appropriate coagulant that can coagulate the colloidal rad-Cs should greatly improve the Cs removal.

References

1. United Nations Scientific Committee on the Effects of Atomic Radiation (2000) Report to the general assembly, with scientific annexes Annex C Exposures to the public from man-made sources of radiation. http://www.unscear.org/docs/reports/annexc.pdf. Accessed 30 Mar 2015
2. Ministry of Education, Culture, Sports, Science and Technology (2011) Map of soil contamination with radioactive cesium. http://www.mext.go.jp/b_menu/shingi/chousa/gijyutu/017/shiryo/__icsFiles/afieldfile/2011/09/02/1310688_2.pdf. Accessed 11 Apr 2014
3. Ministry of Land, Infrastructure, Transport and Tourism (2012) Water resources in Japan as of fiscal year 2012. http://www.mlit.go.jp/tochimizushigen/mizsei/hakusyo/H24/2-9.pdf. Accessed 11 Apr 2014
4. Fujikawa Y (2011) Action of the engineers and scientists from Osaka and Kyoto in response to the accident in the Fukushima Daiichi nuclear power plant. Jpn J Health Phys 46(3):244–245
5. Fujikawa Y, Wei P, Fujinaga A, Tsuno H, Ozaki H, Kimura S (2013) Removal of cesium from the extract of municipal water treatment sludges by precipitation with ferrocyanide solids. In: Proceedings of the 15th international conference on environmental remediation and radioactive waste management, Belgium, 8–12 Sept 2013, ICEM2013-96320
6. Fujikawa Y, Ozaki H, Tsuno H, Wei P, Fujinaga A, Takanami R, Taniguchi S, Kimura S, Giri RR, Lewtas P (2014) Volume reduction of municipal solid wastescontaminated with

radioactive cesium by ferrocyanide coprecipitation technique. In: Nakajima K (ed) Nuclear back-end and transmutation technology for waste disposal: beyond the Fukushima accident. Springer, pp 329–341

7. Ministry of the Environment (2012) Solid waste management and recycling technology of Japan – toward a sustainable society. http://www.env.go.jp/recycle/circul/venous_industry/en/brochure.pdf. Accessed 30 Mar 2015

8. International Atomic Energy Agency (2003) Application of thermal technologies for processing radioactive waste. IAEA Tecdoc 1527. International Atomic Energy Agency, Vienna

9. Sakai S (1996) Municipal solid waste management in Japan. Waste Manag 16(4):95–405

10. Nabeshima Y (1996) Summary of research on waste minimization studies by Japan Waste Research Foundation (JWRF). Waste Manag 16(4):407–415

11. Katsuura H, Inoue T, Hiraoka M, Sakai S (1996) Full-scale plant study on fly ash treatment by the acid extraction process. Waste Manag 16(4):491–499

12. Kim SY, Matsuto T, Tanaka N (2003) Evaluation of pre-treatment methods for landfill disposal of residues from municipal solid waste incineration. Waste Manag Res 21:416–423

13. Furuta H, Matsumoto S, Higuchi S, Misumi F, Hashimoto T, Nakamura H, Horii Y (2006) Feasibility studies for practical use of the WOW system. APLAS, Shanghai, pp 252–260

14. Parajuli D, Tanaka H, Hakuta Y, Minami K, Fukuda S, Umeoka K, Kamimura R, Hayashi Y, Ouchi M, Kawamoto T (2013) Dealing with the aftermath of Fukushima Daiichi nuclear accident: decontamination of radioactive cesium enriched ash. Environ Sci Technol 47:3800–38706

15. Saffaradeh A, Shimaoka T, Kakuta Y, Kawano T (2014) Cesium distribution and phases in proxy experiments on the incineration od radioactively contaminated waste from Fukushima area. J Environ Radioact 136:76–84

16. Kozai N, Suzuki S, Aoyagi N, Sakamoto F, Ohnuki T (2015) Radioactive fallout cesium in sewage sludge ash produced after the Fukushima Daiichi nuclear accident. Water Res 68:616–626

17. Technical Advisory Council on Remediation and Waste Management (2013) Technical information of the treatment and final disposal of incinerated ash. http://tacrwm.jp/03_techinfo/pdf/03_03/03_03_01_report_pub.pdf, (in Japanese)

18. National Institute for Environmental Studies (2014) Technical information on washing of fly ash. http://www.nies.go.jp/shinsai/flyashwash_2014.6.pdf, (in Japanese)

19. Haas PA (1993) A review of information on ferrocyanide solids for removal of cesium from solutions. Sep Sci Technol 28(17&18):2479–2506

20. Loos-Neskovic C, Fedoroff M (1989) Fixation mechanisms of cesium on nickel and zinc ferrocyanides. Solvent Extr Ion Exch 7(1):131–158

21. Barton GB, Hepworth JL, McClannaham ED Jr, Moore RL, Van Tuyl HH (1958) Chemical processing wastes: recovering fission products. Ind Eng Chem 50(2):212–216

22. Fujikawa Y, Wei P, Tsuno H, Ozaki H, Fujinaga A, Tanitguchi S, Takanami R, Sakurai S (2014) Waste volume reduction ratio when ferrocyanide coprecipitation technique was applied for decontamination of solid waste contaminated with radioactive cesium. In: 69th meeting of Japan Society of Civil Engineers, 10–12 Sept 2014, Osaka, (in Japanese)

23. Nishizaki Y, Miyamae H, Takano T, Izumiya K, Kumagai N (2012) Removal of radioactive cesium from molten fly ash. J Environ Conserv Eng 41(9):569–574

24. Ministry of Land, Infrastructure, Transport and Tourism (2011) Measures against radioactivity in the sewage system, outline of the study and the outcome. http://www.mlit.go.jp/common/000213235.pdf. Accessed 15 Jun 2013

25. Stumm W, Morgan JJ (1981) Aquatic chemistry, Chap. 13. Wiley Interscience, New York

26. Patnaik P (2004) Dean's analytical chemistry handbook, Chap. 4, 2nd edn. McGraw-Hill, New York

Part III
Environmental Radiation and External Exposure

Chapter 10
Development and Operation of a Carborne Survey System, KURAMA

Minoru Tanigaki

Abstract A carborne survey system named as KURAMA (Kyoto University RAdiation MApping system) has been developed as a response to the nuclear accident at TEPCO Fukushima Daiichi Nuclear Power Plant. KURAMA is a γ-ray survey system with the global positioning system (GPS) and up-to-date network technologies developed for a primary use of carborne surveys. Based on the success of KURAMA, KURAMA-II, an improved version of KURAMA with better handling and ruggedness, is developed for the autonomous operation in public vehicles to minimize the workload of long-standing radiation monitoring required. Around two hundreds of KURAMA-II now serve for the continuous monitoring in residential areas by local buses as well as the periodical monitoring in Eastern Japan by the Japanese government. The outline and present status of KURAMA and KURAMA-II are introduced.

Keywords Radiometry • Mapping • γ-ray • Carborne survey • Air dose rate • Fukushima Daiichi nuclear power plant

10.1 Introduction

The magnitude-9 earthquake in Eastern Japan on 11 March 2011 and the following massive tsunami caused the serious nuclear disaster of Fukushima Daiichi Nuclear Power Plant, which Japan had never experienced before. Huge amounts of radioactive isotopes were released in Fukushima and surrounding prefectures.

In such nuclear disasters, air dose rate maps are quite important to take measures to deal with the incident, such as assessing the radiological dose to the public, making plans for minimizing exposure to the public, and establishing procedures for environmental reclamation. The carborne γ-ray survey technique is known to be one of the effective methods to make air dose rate maps [2]. In this technique,

M. Tanigaki (✉)
Research Reactor Institute, Kyoto University, 2-1010 Asashironishi, Kumatori, Osaka 590-0494, Japan
e-mail: tanigaki@rri.kyoto-u.ac.jp

© The Author(s) 2016
T. Takahashi (ed.), *Radiological Issues for Fukushima's Revitalized Future*,
DOI 10.1007/978-4-431-55848-4_10

a continuous radiation measurement with location data throughout the subject area is performed by one or more monitoring cars equipped with radiation detectors. Unfortunately, the existing monitoring system didn't work well in the incident. Such monitoring cars tend to be multifunctional, thus too expensive to own multiple monitoring cars in a prefecture. Fukushima was the case, and to their worse, the only monitoring car and the data center were contaminated by radioactive materials released by the hydrogen explosions of the nuclear power plant. The monitoring cars owned by other prefectures were then collected, but such monitoring cars were too heavy to drive on heavily damaged roads in Fukushima. Then daily measurements of the air dose rate in the whole area of Fukushima were eventually performed by humans. The measuring personnel drove around more than 50 fixed points in Fukushima prefecture twice a day, and they measured the air dose rate of each point by portable survey meters. Airborne γ-ray surveys were performed by the Ministry of Education, Culture, Sports, Science and Technology of Japan (MEXT) and the US Department of Energy, but difficulties in the arrangement of aircraft, aviation regulations, and their flight schedules prevented immediate and frequent surveys in the areas of interest.

KURAMA was developed to overcome such difficulties in radiation surveys and to establish air dose-rate maps during the present incident. KURAMA was designed based on consumer products, enabling sufficient numbers of in-vehicle apparatus to be prepared within a short period. KURAMA realized high flexibility in the configuration of data processing hubs or in the arrangement of monitoring cars with the help of cloud technology. Based on the success of KURAMA, KURAMA-II was developed to realize the continuous monitoring in residential areas. An outline of KURAMA and KURAMA-II and their applications are presented.

10.2 KURAMA

KURAMA [10] is a γ-ray survey system with the global positioning system (GPS) and up-to-date network technologies developed for a primary use of carborne surveys. The system outline of KURAMA is shown in Fig. 10.1.

An in-vehicle unit of KURAMA consists of a conventional NaI scintillation survey meter with an appropriate energy compensation, an interface box for the analog voltage output from the detector to a USB port of PC, a GPS unit, a laptop PC, and a mobile Wi-Fi router (Fig. 10.2). Its simple and compact configuration allows users to set up an in-vehicle unit in a common automobile. The software of in-vehicle part is developed with LabVIEW. The radiation data collected every 3 s is tagged by its respective location data obtained by GPS and stored in a csv file. This csv files updated by respective monitoring cars are simultaneously shared with remote servers by Dropbox over a 3G network, unlike other typical carborne survey systems in which special telemetry systems or storage media are used for data

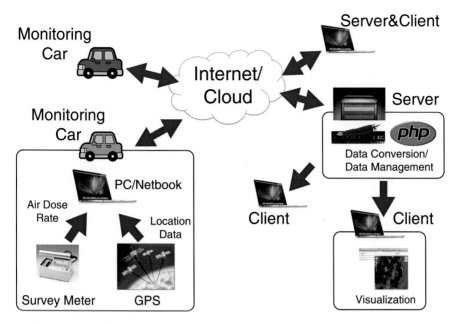

Fig. 10.1 The system outline of KURAMA. Monitoring cars and servers are connected over the Internet by cloud technology

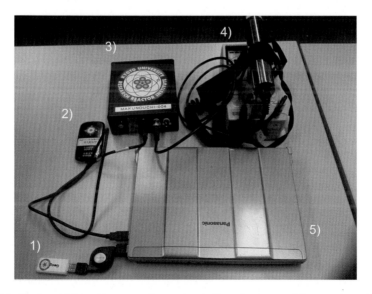

Fig. 10.2 The in-vehicle part is compactly composed of mostly commercial components. (*1*) GPS unit, (*2*) 3G mobile Wi-Fi router, (*3*) MAKUNOUCHI, (*4*) NaI survey meter, and (*5*) PC

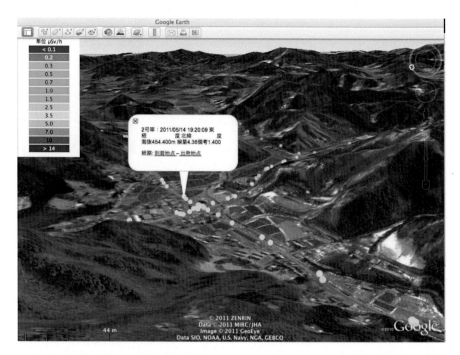

Fig. 10.3 The data is simultaneously plotted on Google Earth. The color of each dot represents the air dose rate at respective point

collection. With this feature, anyone can set up their own "data center" anywhere as far as a conventional Internet connection and a PC with Dropbox are available. This kind of flexibility should be required in disasters like the present case because the carborne survey system owned by Fukushima prefecture eventually came to a halt due to the shutdown of the data center by the disaster.

Once the radiation data in csv format is shared with remote servers, the data file is processed by servers in various ways, including the real-time display on Google Earth in client PCs (Fig. 10.3).

10.3 KURAMA-II

Long-term (several tens years) and detailed surveillance of radiations are required in residential areas that are exposed to radioactive materials. Such monitoring can be realized if moving vehicles in residential areas such as buses, delivery vans, or bikes for mail delivery have KURAMA onboard. KURAMA-II [11] is designed for such purpose.

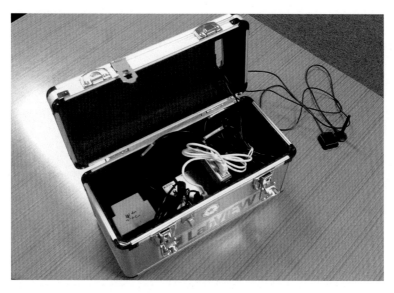

Fig. 10.4 The in-vehicle unit of KURAMA-II. A CsI detector and a CompactRIO are compactly placed in a toolbox with the size of 34.5 × 17.5 × 19.5 cm

KURAMA-II stands on the architecture of KURAMA, but the in-vehicle part is totally redesigned. The platform is based on CompactRIO series by National Instruments to obtain better toughness, stability, and compactness. The radiation detection part is replaced from the conventional NaI survey meter to a Hamamatsu C12137 detector [4], a CsI detector characterized by its compactness, high efficiency, direct ADC output, and USB power operation. The direct ADC output enables to obtain γ-ray energy spectra during operation. The mobile network and GPS functions are handled by a Gxxx 3G series module for CompactRIO by SEA [5]. All of the components for the in-vehicle part are placed in a small toolbox for a better handling (Fig. 10.4).

The software for KURAMA-II is basically the same code as that of original KURAMA, thanks to the good compatibility of LabVIEW over various platforms. Additional developments were performed in several components such as device control software for newly introduced C12137 detector and Gxxx 3G module, the start-up and initialization sequences for autonomous operation, and the file transfer protocol.

CompactRIO is designed for applications in harsh environment and limited space. Therefore, KURAMA-II can be used other than carborne surveys (Fig. 10.5). For example, KURAMA-II is loaded on a motorcycle intending not only the attachment with motorcycles for mail delivery, but also the monitoring in regions where conventional cars cannot be driven, such as small paths between rice fields or those through forests. Also, KURAMA-II with DGPS unit is prepared for the precise mapping by walking in rice fields, orchards, parks, and playgrounds in Fukushima.

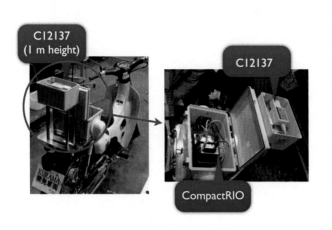

Fig. 10.5 KURAMA-II for bike survey (*left* and *middle*) and walking survey (*right*). All have basically the same hardware and software configuration with different ways of installation. In the case of walking survey, the existing GPS part is replaced with DGPS for the better precision of positioning measurement

10.4 Applications of KURAMA and KURAMA-II

As the developer of KURAMA and KURAMA-II, we have demonstrated possible applications of KURAMA and KURAMA-II through a series of field tests. In the beginning, we demonstrated an efficient γ-ray carborne survey by KURAMA in Fukushima prefecture in collaboration with the Fukushima prefectural government (Fig. 10.6). This result encourages the Ministry of Education, Culture, Sports, Science and Technology in Japan (MEXT) to conduct the first official carborne survey project by KURAMA.

We then carried out a field test of continuous monitoring by KURAMA-II on a local bus in Fukushima city in December 2011 in collaboration with Fukushima Kotsu Co. Ltd., one of the largest bus operators in Fukushima prefecture (Fig. 10.7). Local buses are suitable for continuous monitoring purpose because of their fixed routes in the center of residential areas and routine operations.

Based on the success of the field test on a local bus in Fukushima city, the region of this field test has been extended to other major cities in Fukushima prefecture since January 2013, i.e., Koriyama city, Iwaki city, and Aizuwakamatsu city. Five KURAMA-II in-vehicle units are deployed for this test, and the result is summarized and released to the public from the website [6] on a weekly basis.

The team of the Fukushima prefectural government made precise radiation maps of major cities in Fukushima prefecture mainly for "hot spot" search just after KURAMA was available [3]. Soon, the Fukushima prefectural government sought the possibility to extend the radiation monitoring by local buses over Fukushima

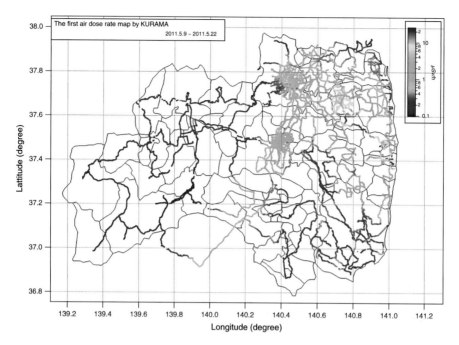

Fig. 10.6 The air dose rate map generated by the first demonstration of KURAMA in May 2011. This was also the first prefecture-wide map of air dose rate in this accident

Fig. 10.7 KURAMA-II under the field test on a local bus

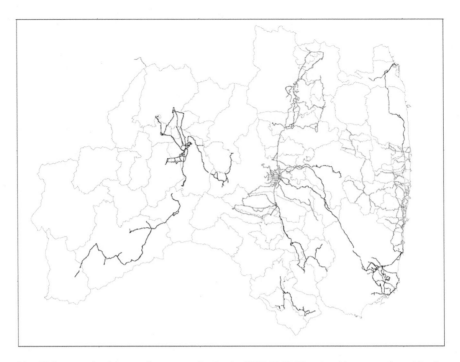

Fig. 10.8 A result of the continuous monitoring by KURAMA-II on local buses conducted by the Fukushima prefectural government [1]

prefecture because they realized the importance of continuous monitoring in residential areas. In August 2015, the monitoring scheme by local buses started as an official operation by the Fukushima prefectural government with the collaboration of Kyoto University and Japan Atomic Energy Agency (JAEA). As of May 2015, 30 local buses owned by bus companies and 20 official cars owned by Fukushima prefecture are continuously operated throughout Fukushima prefecture. Real-time data is released to the public on the display system at the public space of a building in Fukushima city, and the summarized results are available on a weekly basis on the web (Fig. 10.8) [1].

MEXT conducted the very first official carborne survey in Fukushima prefecture and its surrounding area in June 2011. Then they extended carborne surveys in Eastern Japan [8] including Tokyo metropolitan area in December 2011. MEXT started a carborne survey project in March 2012, in which 100 units of KURAMA-II were lent to municipalities in Eastern Japan [7]. KURAMA-II was placed in sedan cars of municipalities, and the cars were driven around by ordinary staff members in respective municipalities, who didn't have any special training on radiation measurement. This survey was successful and proved the performance and scalability of KURAMA-II system. Now, this survey has been conducted periodically by MEXT and the national regulation authority (NRA), which is the successor of the radiation monitoring of the present incident (Fig. 10.9).

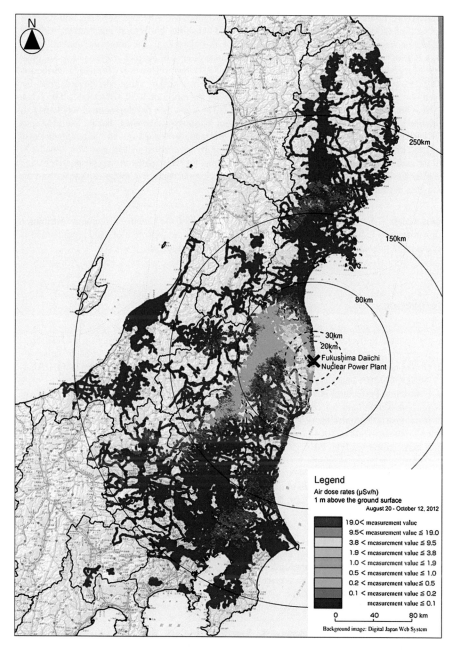

Fig. 10.9 Map of air dose rates on roads measured by KURAMA-II in the periodical survey conducted by NRA between August and October 2012 [9]

Acknowledgements The author is grateful to Dr. Mizuno, Mr. Abe, Mr. Koyama, and the staff members of the KURAMA operation team at the Fukushima prefectural government for their continuous support to the field tests of KURAMA in Fukushima. The author is thankful to Dr. Ito of ICR, Kyoto University; Prof. Maeno of Graduate School of Science, Kyoto University; and the staff members of the KURAMA field test team of RRI, Kyoto University for their contribution to the test operation in Fukushima. The author is indebted to Dr. Saito, Mr. Yoshida, and Dr. Takemiya at JAEA for the discussions concerning the operation of KURAMA and to Dr. Tsuda at JAEA for evaluating the $G(E)$ functions of C12137 series. The development of KURAMA-II is adopted by "Japan recovery grant program" from National Instruments, Japan. The field tests of KURAMA-II on local buses are supported by Fukushima Kotsu Co. Ltd., Shin Joban Kotsu Co. Ltd., and Aizu Noriai Jidosha Co. Ltd. The author is indebted to Mr. and Mrs. Takahashi and the staff members at Matsushimaya Inn, Fukushima, for their heartwarming hospitality during the activities in Fukushima regardless of their severe circumstances due to the earthquake and the following nuclear accident.

References

1. Fukushima Prefectural Government (2014) Vehicle-borne survey monitoring in fukushima prefecture, Jan 2014. https://www.pref.fukushima.lg.jp/sec/16025d/soukou.html (in Japanese)
2. Guidelines for Radioelement Mapping Using Gamma Ray Spectrometry Data, IAEA-TECDOC-1363, International Atomic Energy Agency, 2003, p 40
3. Fukushima Prefectural government (2011) Air dose measurement by KURAMA. http://www.pref.fukushima.lg.jp/sec/16025c/genan28.html (in Japanese)
4. Hamamatsu Photonics Inc (2011) Radiation detection module. http://www.hamamatsu.com/jp/en/product/category/3100/4012/4134/C12137/index.html
5. S.E.A. gmbh (2011) SEA 9724 product flyer. http://www.sea-gmbh.com/en/products/compactrio-products/mobile-communications/sea-9724-3ggps/
6. Research Reactor Institute, Kyoto University (2012) Field test of KURAMA-II. http://www.rri.kyoto-u.ac.jp/kurama/kouiki/kurama2_test.html (in Japanese)
7. Ministry of Education, Culture, Sports, Science and Technology (2012) Results of continuous measurement of air dose rates through a vehicle-borne survey (as of March 2012). http://radioactivity.nsr.go.jp/en/contents/6000/5637/24/338_Suv_091218_e.pdf
8. Press Release (2012) Results of continuous measurement of air dose rates through a vehicle-borne survey by MEXT (as of Dec 2011), 21 Mar 2012, MEXT. http://radioactivity.mext.go.jp/en/contents/5000/4688/view.html
9. Saito K (2013) Studies on the precise distribution and trends of air dose rate on the road by using carborne survey technique – the introduction of the progress of the third distribution study, Feb 2013. http://www.jaea.go.jp/fukushima/kankyoanzen/tyouki-eikyou/giji/08/pdf/8-2-1.pdf (in Japanese)
10. Tanigaki M, Okumura R, Takamiya K, Sato N, Yoshino H, Yamana H (2013) Development of a carborne γ-ray survey system, KURAMA. Nucl Instrum Methods Phys Res Sect A 726: 162–168
11. Tanigaki M, Okumura R, Takamiya K, Sato N, Yoshino H, Yoshinaga H, Kobayashi Y, Uehara U, Yamana H (2015) Development of KURAMA-II and its operation in Fukushima. Nucl Instrum Methods Phys Res Sect A 781:57–64

Chapter 11
In Situ Environmental Radioactivity Measurement in High–Dose Rate Areas Using a CdZnTe Semiconductor Detector

Munehiko Kowatari, Takumi Kubota, Yuji Shibahara, Toshiyuki Fujii, Koichi Takamiya, Satoru Mizuno, and Hajimu Yamana

Abstract For the purpose of determining a surface deposition density on soil for radio-cesiums, a CdZnTe (CZT) semiconductor detector whose crystal has dimensions of 1 cm cubic was applied to the in situ environmental radioactivity measurement in deeply contaminated areas in Fukushima region. Even in high–dose rate areas where pulse height spectra weren't able to be properly obtained by the conventional high-purity Ge (Hp-Ge) semiconductor detector, proper pulse height spectra were obtained by the CZT detector with certain accuracy. Results of deposition density on soil for ^{134}Cs and ^{137}Cs derived from net peak areas by the CZT detector seemed consistent, comparing with those measured by the Japanese government. Air kerma rates were estimated by the same pulse height spectra for determining surface deposition density on soil for radio-cesiums. Estimated results showed almost the same values as obtained by the NaI(Tl) scintillation survey meter. The results indicate that the CZT detector can be applied to rapid and simple in situ gamma ray radioactivity measurement in higher–dose rate areas whose dose rates exceed several tenth μSv h^{-1}. The study also strongly supports that the CZT detector is one promising candidate for the detector to be used for checking the effect of decontamination works and for long-term monitoring in heavily contaminated areas.

Keywords CdZnTe detector • In situ gamma ray environmental measurement • Decontamination • Dose rate • Environmental radiation monitoring • Surface deposition density on soil • ^{137}Cs • ^{134}Cs

M. Kowatari (✉)
Department of Radiation Protection, Nuclear Science Research Institute, Japan Atomic Energy Agency, 2-4 Shirakata, Tokai, Naka, Ibaraki 319-1195, Japan
e-mail: kowatari.munehiko@jaea.go.jp

T. Kubota • Y. Shibahara • T. Fujii • K. Takamiya • H. Yamana
Research Reactor Institute, Kyoto University, Kumatori, Osaka 590-0494, Japan

S. Mizuno
Nuclear Power Safety Division, Fukushima Prefectural Government, Fukushima, Fukushima 960-8670, Japan

© The Author(s) 2016
T. Takahashi (ed.), *Radiological Issues for Fukushima's Revitalized Future*,
DOI 10.1007/978-4-431-55848-4_11

11.1   Introduction

After the Fukushima Nuclear Accident, contamination by a vast amount of radioactive materials released due the accident still remains in areas close to the Fukushima Daiichi Nuclear Power Station (FDNPS). Decontamination and remediation works in heavily contaminated areas are recognized a challenging practice. Particularly, methodologies of decontamination and its confirmation of effect in deeply contaminated areas where there are consequently high dose rates have been being intensively developed.

Among methodologies for determining the surface deposition density on soil, in situ environmental gamma ray measurement using high-purity Ge detector [1, 2] is widely recognized effective and reliable and has been playing an important role in the measurement campaign in the whole region of northern part of Japan under the conduction of the government of Japan (http://radioactivity.nsr.go.jp/en/). On the other hand, high dose rates more than tenth μSv h^{-1} impede the precise spectrometry using the Ge detectors caused by their own high sensitivity to gamma rays. In our measurement campaign for investigating the isotropic ratio of radio-cesiums [3, 4], the Ge detector used was not able to show its high energy resolution when measured in higher–dose rate areas. This was caused by broadening each peak in the pulse height spectrum due to overlapping of too many events incoming the detector sensitive region.

Comparing with Ge detectors, a newly developed CdZnTe (CZT) detector has less energy resolution of pulse height spectrum. A CZT detector which has a small detector element is also less sensitive to gamma rays from radio-cesiums. However, a CZT detector has enough energy resolution to distinguish peaks by gamma rays from ^{134}Cs and ^{137}Cs (i.e. its progeny, ^{137m}Ba). The lower sensitivity to gamma rays would enable us to appropriately measure pulse height spectra in the environment whose dose rates reach more than few tenth μSv h^{-1}. In addition, a CZT detector is an easy-to-use, light-weight and coolant-free detector. Considering these characteristics, a CZT detector also would help determine environmental radioactivity even in deeply contaminated areas whose dose rates exceed more than a couple of tenth μSv h^{-1}.

This article describes investigation on the adaptability of a CZT detector to an in situ environmental radioactivity measurement. Before measurements, energy dependence of peak efficiency and angular dependence of peak efficiency for the CZT detector were evaluated. A series of environmental radioactivity measurements were then conducted within the region within several kilometers in radius centering the FDNPS, which is designated as the so-called difficult-to-return zone, which is deeply contaminated areas due to radio-cesiums from the FDNPS. There are still some places whose ambient dose equivalent rates exceed several tenth μSv h^{-1} and require to be remediated. Surface contamination densities on soil due to radio-cesiums were determined from measured pulse height spectra. Air kerma rates at measurement points were also estimated from the same pulse height spectra by means of three methodologies. Measured results were then compared with those obtained by the government of Japan (http://radioactivity.nsr.go.jp/en/). The usability of the CZT was also discussed in this article.

11.2 Materials and Methods

In this study, a CZT detector was used as a detector which applies the rapid and simple in situ environmental gamma ray measurement for determining surface deposition on soil due to radio-cesiums. Considering its characteristics of gamma ray detection described above, a CZT detector is taken as one promising candidate for the purpose. In addition, high-dose areas limit remediation and monitoring activities. As summarized in Table 11.1, a light, easy-to-use, USB-powered and coolant-free CZT detector allows us to expand our remediation and monitoring activities. This strongly will promise rapid, safe and efficient monitoring activities for a limited time. Figure 11.1 shows the appearance and cross-sectional drawing for the CZT detector assembly (Kromek GR1^{TM}) used in this study. The sensitive region of the detector is made of CdZnTe semiconductor crystal, and the dimension of the sensitive region is 1 cm^3.

Surface deposition densities of radio-cesiums using the CZT detector were measured from 27 to 28 May 2013 and from 11 to 13 December 2013. The CZT detector was set at the height of 1 m, using a conventional tripod. For obtaining pulse height spectra from the CZT detector and for power supply, a laptop PC was connected to the detector by a USB cable. Open and level fields which have no buildings within at least 10 m from the center of measuring points were selected, as possible. Ambient dose equivalent rates at measurement points were

Table 11.1 Comparison of characteristics of the conventional Ge semiconductor detector and the CZT detector

	High purity Ge semiconductor detector	CdZnTe semiconductor detector
Detector element	Ge	CdZnTe
Typical dimension and shape of detector element	Cylindrical 5.0 cm $\varphi \times$ 6.0 cm^L	Cubic $1.0 \times 1.0 \times 1.0$ cm^3
Typical energy resolution for ^{137}Cs: 662 keV (FWHM) (%)	$0.2 \sim 0.3$	$2.0 \sim 2.5$
Weight (kg)	$\sim$10 (including LN$_2$)	0.05
Cost (Japanese yen) (initial, including measurement system)	$3,000,000 \sim 10,000,000$	$600,000 \sim 1,000,000$
(running)	$10,000 \sim$ per 1 day	$0 \sim$
Necessary to be cooled	Yes (LN$_2$ or electric cooling system)	None
External power supply	Required	None

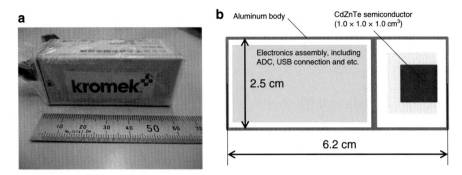

Fig. 11.1 Appearance (**a**) and cross sectional drawing (**b**) for the CZT detector assembly (Kromek GR1™)

also monitored using the NaI(Tl) scintillation survey meter and ranging between 0.2 and 45 μSv h^{-1}. Measurement time for each point changed according to the dose rates and was set to 600 s for high–dose rate areas and 1200 s for lower–dose rate areas, respectively.

Surface deposition densities on soil of each radio-cesium A_a (kBq m^{-2}) were determined by dividing peak areas per second N_f (cps) due to gamma rays from ^{134}Cs and ^{137}Cs by peak efficiencies with energies of gamma rays, P_{eff} (cps (kBq m^{-2})$^{-1}$). Equation (11.1) shows the surface deposition density on soil, A_a:

$$A_a = \frac{N_f}{P_{eff}} \qquad (11.1)$$

The peak efficiency, P_{eff}, for an incident gamma ray with energy of E (keV) was derived in accordance with Eq. (11.2):

$$P_{eff} = \frac{N_0}{\phi} \frac{N_f}{N_0} \frac{\phi}{A_a} \qquad (11.2)$$

where N_0/φ is a net peak area for full energy absorption of incident gamma ray per unit gamma ray fluence. N_0/φ, was then obtained by measuring pulse height spectra using checking radioactive sources such as ^{152}Eu, ^{60}Co and ^{137}Cs and by calculating using the EGS4 Monte Carlo code. N_f/N_0 is the correction factor for angular dependence of a net peak area for full energy absorption of the CZT detector used. φ/A_a ((cm^{-2} s^{-1}) (kBq m^{-2})$^{-1}$) is a gamma ray fluence rate at 1 m above the ground per unit surface deposition density of each radionuclide of interest, namely, ^{134}Cs and ^{137}Cs. This depends on the migration of soil from the surface of ground, and depth profile of concentration of radionuclides in soil varies as time elapses from the initial deposition. The parameter defined as a relaxation mass per unit area, β (g cm^{-2}), is used to express the vertical distribution of radionuclides deep inside soil [1, 2]. In this study, β was set to 2 and/or 3, in accordance with the recommendation of reference 2. The reference 2 explains the effect of attenuation of gamma rays due to surface soil and migration of radio nuclides on soil according to the time elapsed from initial deposition of radionuclides to the measurement date in detail.

Air kerma rates at measurement points were estimated from measured surface contamination densities on soil by the CZT detector. In addition, using the same spectra obtained by the CZT detector, air kerma rates at measurement places were estimated by the stripping method [5] and the G(E) function method [6]. Evaluating dose rate using the same pulse height spectra would enhance information on measurement points, help shorten staying time in higher–dose rate areas and facilitate a long-term monitoring of heavily contaminated areas. Regarding the stripping method, the response matrix of the CZT detector to mono-energetic gamma rays ranging from 50 keV to 3 MeV was calculated by MCNP-4C code. The pulse height spectrum was decomposed to gamma ray fluence spectrum by applying calculated response matrix. The air kerma rate was then estimated by multiplying fluence-to-air-kerma conversion coefficient taken from ICRP 74 [7] with obtained gamma ray fluence. In contrast, the G(E) function method enables us to directly determine air kerma rates from the accumulated pulse height spectra, by employing the operator defined, "G(E) function." The method has also taken advantage of direct determination of air kerma rates which the effect of scattered gamma ray components is included. The previous literature gives more detailed explanations on the G(E) function method and its application to environmental radiation measurement [6]. The G(E) function for the CZT detector was derived from the calculated responses matrix of the CZT detector to mono-energetic gamma rays ranging from 50 keV to 3 MeV. The EGS4 was used for the calculation. Before determining air kerma rates by pulse height spectra obtained in the field, the calculated response functions were verified by comparing the measured spectra from point sources of ^{137}Cs with the calculated spectra.

11.3   Results and Discussion

As basic characteristics of the detectors for in situ environmental radioactivity measurement, energy and angular dependences of full energy absorption peak efficiency of the CZT detector were evaluated. Figure 11.2 shows measured and calculated peak area per unit gamma ray fluence as a function of gamma ray energy. In the energy region around 100 keV, peak efficiency reaches up to almost 0.9. On the other hand, quite low values with energies more than around 1500 keV suggest that the CZT detector has less sensitivity to energetic gamma ray from ^{40}K and/or ^{208}Tl. Figure 11.3 shows the angular dependence of a relative net peak area for full energy absorption as a function of incident gamma ray energy. In this figure, 0° corresponds to a direction of the detector axis, and 90° is a horizontal direction to the ground. Values in the Fig. 11.3 were normalized by those obtained at 90°. Figure 11.3 indicates that calculated net peak areas for incident gamma rays

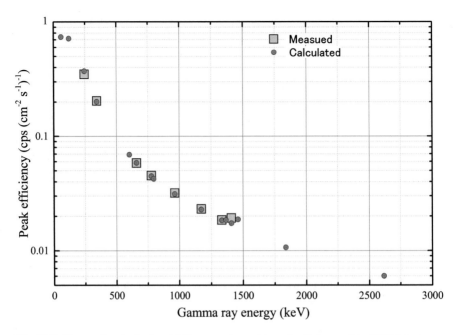

Fig. 11.2 Measured and calculated full energy absorption peak efficiency of the CZT detector

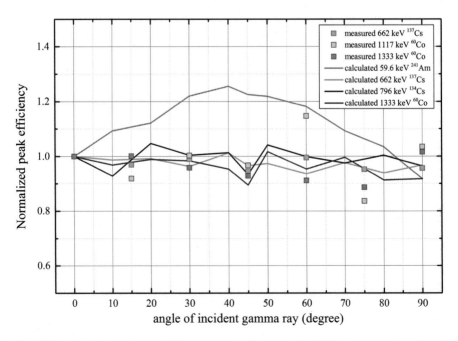

Fig. 11.3 Angular dependence of full energy peak efficiency of the CZT detector as a function of incident gamma ray energy

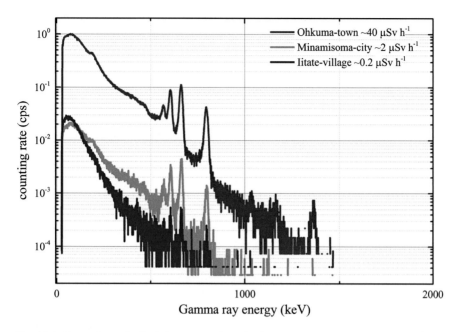

Fig. 11.4 Comparison of pulse height spectra from the CZT detector obtained in areas whose dose rates range between 0.2 and 40 μSv h^{-1}

with energies between 662 and 1333 keV was evaluated to be within 10 % of those obtained at 90° by Monte Carlo calculations using MCNP-4C. Measured results were also found to be within 20 % of those obtained at 90°. This shows that the CZT detector has less angular dependence to gamma rays with energies between 662 and 1333 keV. From measured and calculated results, values of N_f/N_0 for ^{134}Cs and ^{137}Cs gamma rays were both taken as 1.0. Results shown in Figs. 11.1 and 11.2 imply that the CZT detector could allow us to properly measure pulse height spectra in heavily contaminated areas.

In situ environmental radioactivity measurements using the CZT detector were made inside the difficult-to-return zone at several kilometers from the FDNPS and also within 40 km in radius centering the FDNPS. Figure 11.4 shows examples of pulse height spectra obtained in areas whose ambient dose equivalent rates range from 0.2 to 45 μSv h^{-1}. Pulse height spectrum was obtained in 600 s at measurement point whose dose rate was monitored to be 45 μSv h^{-1}. As shown in Fig. 11.4, peaks due to gamma rays from radio-cesiums were clearly identified. The dead time observed in the area whose dose rate was around 40 μSv h^{-1} was around 1 %, and this would lead to the proper measurement for determining on surface deposition density on soil of radio-cesiums. Peaks due to gamma rays from ^{40}K, however, could not be seen in each pulse height spectrum. As shown in Fig. 11.1, the CZT detector has less sensitivity to energetic gamma rays more than around 1500 keV, comparing those from ^{137}Cs. The CZT detector was not able to observe events enough to form the peak due to gamma rays with energy of 1460 keV from ^{40}K.

This might also be rather advantageous, if the detector were to be used for confirming the effect of the decontamination work in higher–dose rate regions. It would be a crucial disadvantage that the CZT detector could not clearly identify peaks due to gamma rays from ^{40}K in the case of general environmental radioactivity measurement. On the other hand, gamma rays emitted from ^{134}Cs and ^{137}Cs have been still contributing to high dose rate in the "difficult-to-return areas." Considering that gamma rays emitted by these contaminants should be well identified, the effect of higher-energy gamma rays from ^{40}K is properly eliminated from pulse height spectra obtained by the CZT detector. From this viewpoint, the CZT detector enables to easily confirm the effect of decontamination work with certain accuracy.

Measured surface deposition densities on soil of ^{134}Cs and ^{137}Cs were summarized in Table 11.2, derived from net peak areas in pulse height spectra and calculated in accordance with Eq. (11.1). All the results in Table 11.2 were corrected to the date of the initial deposition (11 March 2011), by applying half-life of ^{134}Cs (2.0684 y) and ^{137}Cs (30.167 y). In addition to the surface deposition density on soil, air kerma rates due to radio-cesiums at measurement date were estimated from measured results and also listed in the table. Measured results seemed consistent, comparing with those evaluated by the measurement campaign by the government of Japan (http://radioactivity.nsr.go.jp/en/). Results obtained in wider regions are also summarized in our previous report [4].

Table 11.2 Comparison of surface deposition densities on soil within 7 km in radius centering the FDNPS. Measurement was conducted from 27th May 2013 to 28th May 2013

Location	Coordinate of location	Distance from FDNPS (km)	Measured surface deposition density on soil at 11th March, 2011 (kBq m^{-2})		Estimated air kerma rate due to gamma rays from ^{134}Cs and ^{137}Cs at measuring date (μGy h^{-1})
			^{134}Cs	^{137}Cs	
Ohkuma-town, Chuodai	37:25:26 N 141:00:08 E	2.7, W	$(1.10 \pm 0.02) \times 10^4$	$(1.05 \pm 0.03) \times 10^4$	35.8 ± 0.5
Ohkuma-town, Ohno station	37:24:32 N 140:59:07 E	4.4, WSW	$(3.29 \pm 0.24) \times 10^3$	$(3.37 \pm 0.08) \times 10^3$	10.7 ± 0.5
Futaba-town, Kami-hatori	37:27:25 N 140:59:22 E	5.6, NW	$(1.89 \pm 0.13) \times 10^3$	$(1.88 \pm 0.05) \times 10^3$	6.11 ± 0.26

Table 11.3 Comparison of measured dose equivalent rate and estimated air kerma rate in terms of various methods

Location	Coordinate of location	Distance from FDNPS (km)	Ambient dose equivalent rate by the conventional NaI(Tl) scintillation survey meter (μSv h^{-1})	Estimated air kerma rate due to ^{134}Cs and ^{137}Cs at measuring date (μGy h^{-1})	Estimated air kerma rate by the stripping method (μGy h^{-1})	Estimated air kerma rate by the G(E) function method (μGy h^{-1})
Ohkuma-town, Chuodai	37:25:26 N 141:00:08 E	2.7, W	45.3 ± 0.4	35.8 ± 0.5	33.2	37.1
Ohkuma-town, Ohno station	37:24:32 N 140:59:07 E	4.4, WSW	12.9 ± 0.1	10.7 ± 0.5	9.35	14.1
Futaba-town, Kamiha-tori	37:27:25 N 140:59:22 E	5.6, NW	1.63 ± 0.02	6.11 ± 0.26	2.17	3.29

Table 11.3 shows the comparison of air kerma rates, $\dot{K}_{air}$ (μGy h^{-1}), estimated from measured surface contamination densities on soil and in terms of the stripping and G(E) function methods. Measured ambient dose equivalent rates, $\dot{H}^*(10)$ (μSv h^{-1}), at the same measurement points were also listed. All results were corrected at measuring date (28 May 2013). Results obtained in high–dose rate areas were considered to be identical within 6 %. However, results obtained in lower–dose rate region have large discrepancy, particularly between air kerma rate estimated from surface deposition densities on soil and others. This might be caused by improper setting of relaxation mass, β, for determining the surface deposition densities on soil. On the other hand, air kerma rates estimated by the stripping method and the G(E) function method might reproduce actual air kerma rate, comparing with ambient dose equivalent rate obtained by the NaI(Tl) survey meter. Further investigation should be required for more accurate estimation of dose rate at measurement places.

11.4 Summary

A light-weight, easy-to-handle and cooling-free CdZnTe (CZT) semiconductor detector whose crystal has dimensions of 1 cubic cm was applied to the rapid and simple in situ environmental radioactivity measurement in deeply contaminated areas in Fukushima region in 2013. Even in high–dose rate areas more than a couple of tenth μSv h^{-1}, the CZT detector allowed to obtain the proper and fine pulse height spectrum, which is clearly distinguished peaks from gamma rays due to ^{134}Cs and ^{137}Cs. Results of deposition density on soil for ^{134}Cs and ^{137}Cs derived from net peak areas by the CZT detector were evaluated to be between 1.9×10^3 and 1.1×10^4 kBq m^{-2} for each radio-cesium. They were also found to be consistent, comparing with those measured by the Japanese government and obtained in other literature. Air kerma rates were estimated using the same pulse height spectra by the stripping and the G(E) function methods. Estimated results showed almost the same values as obtained by the NaI(Tl) scintillation survey meter, considering the differences of dosimetric quantities.

Throughout performance tests done in this study, the CZT detector has poor energy resolution and less sensitivity to energetic gamma rays with energies above 1500 keV, namely, gamma rays from ^{40}K. However, this means the CZT detector could eliminate interferences, when used in areas whose dose rates are more than a couple of tenth μSv h^{-1} due to contamination by radio-cesiums. The field monitoring test showed the CZT detector is easy to use and helps shorten staying time at measurement points whose dose rates exceed several tenth μSv h^{-1}. The study also clearly indicates that the CZT detector is one promising candidate for the detector to be used for checking the effect of decontamination works and for long-term monitoring in heavily contaminated areas, in order to accelerate implementations of decontamination and remediation works in "difficult-to-return zone."

Acknowledgements The authors thank all staff members of Hotel Matsushimaya Ryokan in Fukushima-city for their heart-warming hospitality and continuing support for our research project.

Funding This research work was supported by the KUR Research Program for Scientific Basis of Nuclear Safety.

References

1. Beck HL, DeCampo J, Gogolak C (1972) In situ Ge(Li) and NaI(Tl) gamma-ray spectrometry for the measurement of environmental radiation, USAEC report HASL-258. USAEC, New York
2. International Commission on Radiation Units and Measurements (1994) Gamma-ray spectrometry in the environment, ICRU report 53. ICRU, Bethesda
3. Shibahara Y, Kubota T, Fujii T et al (2014) Analysis of cesium isotope compositions in environmental samples by thermal ionization mass spectrometry – 1. A preliminary study for source analysis of radioactive contamination in Fukushima prefecture. J Nucl Sci Technol 51(5):575–579
4. Kowatari M, Kubota T, Shibahara Y et al (2015) Application of a CZT detector to in situ environmental radioactivity measurement in the Fukushima area. Radiat Prot Dosimetry. doi:10.1093/rpd/ncv277

5. Kurosawa T, Iwase H, Saito H et al (2014) Field photon energy spectra in Fukushima after the nuclear accident. J Nucl Sci Technol 51(5):730–734
6. Moriuchi S, Miyanaga I (1966) A method of pulse height weighting using the discrimination bias modulation. Health Phys 12(10):1481–1487
7. International Commission on Radiological Protection (ICRP) (1997) Conversion coefficients for use in radiological protection against external radiation, ICRP publication 74 Ann. ICRP 27(4). Pergamon Press, London/New York

Chapter 12
Safety Evaluation of Radiation Dose Rates in Fukushima Nakadori District

Masayoshi Kawai, Michikuni Shimo, and Muneo Morokuzu

Abstract After the TEPCO Fukushima DAIICHI NPP accident, IAEA and ICRP advised accelerating the decontamination work to clean up the living environment of the areas where additional annual radiation exposure doses are beyond 1 mSv per year (i.e., 1 mSv/y) and to diminish radiation worries. However, the advice was not recognized well because it did not contain clearly understandable numerical data. In the present work, the ambient radiation dose rates in the Nakadori district have been investigated to clarify that the doses are lower than 1 mSv/y in the major part where the decontamination was completed. A part of the district and three municipalities in the special decontamination area have doses of 1.0–2.0 mSv/y. The country-averaged annual doses of natural radiation in the world have been evaluated using the basic data taken from the UNSCEAR 2000 report. The result shows that total annual exposure doses containing cesium and natural radiation contributions in Fukushima are 2–4 mSv/y, which are close to the natural radiation doses in Europe. The risk coefficient of the public exposure limits, 1 mSv/y, has also been evaluated to be 4.5×10^{-7} per year. It is lower than that of traffic accidents by two orders of magnitude. These results will be useful to judge how the safety of the Fukushima prefecture is secured.

Keywords Cesium contamination • Decontamination • Annual exposure dose • 1 mSv/y • Risk coefficient • Natural radiation • Nakadori district • Fukushima

M. Kawai (✉)
High Energy Accelerator Research Organization (KEK), Oho, Tsukuba, Ibaraki 305-0801, Japan
e-mail: masayoshi.kawai@kek.jp

M. Shimo
Fujita Health University, Dengakugakubo, Kutsukake-cho, Toyoake, Aichi 470-1192, Japan

M. Morokuzu
(NPO) Public Outreach, 1-1-11 Nezu, Bunkyo-ku, Tokyo 113-0031, Japan

© The Author(s) 2016
T. Takahashi (ed.), *Radiological Issues for Fukushima's Revitalized Future*,
DOI 10.1007/978-4-431-55848-4_12

12.1 Introduction

More than 4 years have passed since the TEPCO Fukushima Daiichi nuclear power plant accident. The accident brought severe contamination by radioactive cesium isotopes in very wide areas of the Fukushima prefecture as well as neighboring prefectures in the Tohoku and Northern Kanto regions. Decontamination work is being done in order to reduce the ambient radiation level in the living space of the areas where the additional annual radiation exposure doses for individuals are beyond 1 mSv per year (i.e., 1 mSv/y). In the area where the work is assigned to the Fukushima prefecture, about 70 % of the decontamination work plan up to FY 2014 (i.e., up to March 2015) has been completed for housing sites [1]. In the special decontamination area where the work is being done by the Japanese government against the regions having a dose below 50 mSv/y, full-scale decontamination has been completed in four municipalities such as Tamura-shi, Kawauchi-mura, Naraha-machi, and Okuma-machi within FY 2014 [2].

As a result, the ambient radiation dose rates were reduced to values less than 0.23 μSv/h in most parts of the decontaminated areas of the Nakadori district and the average radiation dose rate in the former three municipalities was about 0.4 μSv/h in addition to the effect of natural decay of radioactive nuclides and weathering effects by rain and wind within FY. 2014. The local governments of Tamura-shi and Naraha-machi have declared their intent to remove the evacuation order. However, there are in total about 116,000 people who will be forced to evacuate inside the Fukushima prefecture (69,000) and outside (47,000) in March of 2017 [3]. A recent survey about their will to return reports [4] that 37.3 % of the residents moved within the prefecture and 19.8 % outside it and wanted to return to their home town under certain conditions, whereas half of them did not know what they wanted to do or gave no answer. The residents who moved in Fukushima raised the following conditions for returning: completing the decontamination (48.8 %), lifting the evacuation order in addition to decontamination completion (42.7 %), and disappearance of worries regarding radiation exposure (42.4 %). The people who moved outside Fukushima answered as follows: disappearance of worry about radiation exposure (52.2 %), completing the decontamination (45.7 %), as well as insurance of nuclear plant safety in the future (38.9 %). Anyhow, it should be noted that there is a high proportion of strong or vague worry about radiation influence.

Since 2011 international support activities have been energetically performed to accelerate recovery of the eastern region of Japan, especially the Fukushima prefecture by IAEA and ICRP. IAEA has made many technical advisories on the decontamination of the contaminated area, recovery of the town infrastructure, evacuation and return of the residents, as well as safety reinforcement and its examination of nuclear plants, and so on. In the autumn of 2013, the IAEA's international expert Mission Team for Fukushima Remediation Issues issued important advice in order to accelerate the decontamination and people's return that the government should strengthen its efforts to explain to the public that the additional individual dose of 1 mSv/y is a long-term goal [5]. On the other hand, ICRP has given psychological support through the dialogue meeting with the residents of the Fukushima prefecture mainly to discuss radiation problems concerning its influence on health and radiation protection in their daily life. The representative meeting was held on November 3, 2012 in Fukushima-shi. At the meeting, after the residents' presentation on worry about radiation a few members of ICRP answered that the radiation level in Fukushima was close to the natural radiation in their home town [6]. An explanation of "natural radiation" seems to be very instructive and effective for people to become aware of leading a healthy life even under radiation: people are always exposed to radiation from radioactive isotope intake through food, for example, of K40 accumulations to about 4000 Bq in an adult, as well as cosmic rays and gamma rays from soil and radon. However, their suggestions without understandable scientific numerical data have not worked to change the situation of Fukushima as much as IAEA and ICRP expected.

In the present work, the authors have prepared materials to explain the security situation at Fukushima. First they investigated the radiation distributions in the Nakadori district and its neighboring municipalities, and clarified the annual excess exposure doses. They also evaluated the risk of public standard limits of radiation exposure, 1 mSv/y, and country-averaged annual doses of natural radiation in the world. These data should be very useful for residents to recognize that the present status of the Fukushima prefecture is safe and judge how they will live there.

12.2   Radiation Level of Fukushima Nakadori District

The Fukushima Nakadori district is a region comprising the middle third of the Fukushima prefecture. It is sandwiched between the Aizu district to the west and Hamadori to the east. The Nakadori district contains the large cities of Koriyama–shi, local capital Fukushima-shi, and many middle-sized cities such as Date-shi, Nihonmatsu-shi, and Shirakawa-shi, among others. It occupies a major part in the government and economy including industrial activities and agriculture of the Fukushima prefecture. After the accident, many residents moved here from the Hamadori district having the TEPCO Fukushima NPPs. In Fukushima, contamination by radioactive cesium generally becomes lower with moving to the west therefore the problem of radiation in the Nakadori district is lower than in Hamadori.

Decontamination in the "Intensive Contamination Survey Area" where an additional annual exposure dose between 1 mSv and 20 mSv is promoted for living space by the municipalities, initiating from the higher radiation level zone. The present status of the decontamination work (to March 31, 2015) is shown in Table 12.1 [1]. About 70 % of housing and 97 % of farmland of the implementation plan up to FY 2014 has been decontaminated. In the "Special Decontamination Area" containing a total of 11 municipalities, which consists of the "restricted areas" located within a 20 km radius from TEPCO's Fukushima Daiichi NPP, and "Deliberate Evacuation Areas" where the dose was anticipated to exceed 20 mSv/y, the national government performs the decontamination work except where the radiation level area is higher

Table 12.1 Progress of decontamination work in designated municipalities to March 31, 2015

Item	Housing (Household)	Public facilities (Number of facilities)	Roads (km)	Farmland (ha)
Planned to FY 2014	318,392	8,298	8,572	29,720
Implementations	215,126	6,782	3,767	22,412
Progress rate	67.6 %	81.7 %	43.9 %	75.4 %

than 50 mSv/y, where residents will face difficulties in returning for a long time (i.e., "Difficult-to-Return Zone") and has finished the full-scale decontamination for housing, public facilities, roads, and farmland in Tamura-shi, Naraha-machi, Kawauchi-mura, and Okuma-machi within FY 2014 [2]. Lifting the evacuation order for the former two has been declared by the local government, taking account of a radiation level reduction below 0.4 μSv/h on average, as well as that for the special places of Date-shi, Kawauchi-mura, and Minamisoma-shi where the evacuation orders were spotty and issued from the special decontamination area. Housing decontamination has been completed in Katsurao-mura, Kawamata-machi, and Iitate-mura within FY 2014. Full-scale decontamination is continuing, aiming at completion in FY 2017 or FY 2018 for the remaining seven municipalities.

Owing to decontamination as well as the natural decay of radioactive cesium isotopes and weathering effects, present ambient radiation levels have become fairly lower. Figure 12.1 shows the monitoring information of the environmental radiation dose rate estimated at the 1 m height from the surface of the ground which was measured from an airplane [7]. It is found from the figure that the major part of the Nakadori district is colored by radiation dose rates lower than 0.5 μSv/h, whereas a part of Date-shi and Nihonmatsu-shi has a higher radiation level between 0.5 and 1 μSv/h which seems to be assigned to mountainous zones where decontamination has not been done because they lie outside living spaces.

Table 12.2 compares the radiation dose rates recently measured using a survey meter or a portable type radiation detector located at the representative monitoring posts in the important municipalities in the Fukushima prefecture with those on April 29 of 2011 [8]. The recent data become smaller by a factor of 3 through 12 from those of 2011. A small reducing factor means nondecontamination work. All values except for the Fire Center at Yamakiya in Kawamata-machi are lower than 0.23 μSv/h, which is the long -term target for decontamination.

From the figure and table, it can be said that radiation levels in the living space of the Nakadori district are generally below 0.23 μSv/h so that the additional annual dose will be expected to be below 1 mSv/y. Even at a higher radiation level where the ambient dose rate is 0.5 μSv/h, the additional dose would be 2.5 mSv/y.

Soma-shi, Iwaki–shi, and Hirono-machi in the Hamadori district are in the same situation as Fukushima-shi, Koriyama-shi, and Shirakawa-shi. In the special decontamination area, the average radiation dose rates measured in the housing sites of Tamura-shi, Naraha-machi, and Kawauchi-mura after the decontamination were reported [9] to be 0.35 μSv/h (June 2014), 0.38 μSv/h (June 2014), and

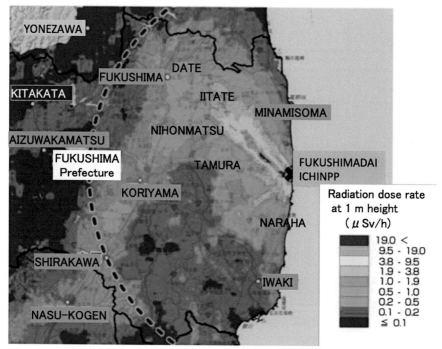

44 months after NPP Accident(2014.11.7)

Fig. 12.1 Monitoring information of environmental radiation dose rates in the Nakadori and Hamadori districts measured from an airplane.

0.41 μSv/h (August 2014), respectively. At the places higher than 1 μSv/h before decontamination, the individual average values were reduced from 1.19 to 0.54 μSv/h (7 %), 1.38 to 0.63 μSv/h (19 %,) and 2.02 to 1.03 μSv/h (30 %) arranged in the same order. The value in parentheses expresses a fraction of the number of housing sites over 1 μSv/h.

12.3   International Support Activities

12.3.1   IAEA's International Expert Mission Team for Fukushima Remediation Issues [5]

In the autumn of 2013, the IAEA's international expert Mission Team for Fukushima Remediation Issues came to Japan in response to the request made by the government of Japan, to follow up with the main purpose of evaluating the progress of the on-going remediation work achieved since the previous mission in October 2011. The Team reported [5] that Japan is allocating enormous resources

Table 12.2 Ambient radiation dose rate at the monitoring posts in Fukushima prefecture (unit: μSv/h)

Place		April 29, 2011	Nov. 7, 2014	April 19, 2015
Date-shi	Government office	1.25, 1.23	0.23, 0.21	0.22, 0.19
Fukushima-shi	Kenpoku health office	1.58	0.24	0.22
Kawamata-machi	Government office	0.73, 0.75	0.16, 0.16	(0.14, 0.14)[a]
Kawamata-machi	Yamakiya fire center	3.0 (approx.)	0.68, 0.68	(0.67, 0.67)[a]
Nihonmatsu-shi	Government office	1.39, 1.44	0.25, 0.26	(0.23, 0.23)[a]
Tamura-shi	Local office at Tokiwa	0.26	0.10, 0.09	(0.10, 0.09)[a]
Koriyama-shi	Common building	1.53	0.13	0.13
Shirakawa-shi	Common building	0.64	0.09	0.09
Aizuwakamatsu-shi	Common building	0.18	0.06	0.04
Minamisoma-shi	Common building	0.54	0.11	0.11
Hirono-machi	Shimokitasako meeting place	0.8 (approx.)	0.11, 0.11	(0.08)[a]
Iwaki-shi	Common building	0.27	0.07	0.07

N.B. [a]Value on March 31, 2015

to developing strategies and plans and implementing remediation activities, with the aim of enhancing the living conditions of the people affected by the nuclear accident, including enabling evacuated people to return and that, as result of these efforts, Japan has achieved good progress in the remediation activities and, in general, has well considered the advice provided by the previous mission in 2011.

It also noted that based on the basic principles of the Act of Special Measures, a system has been established to give priority to remediation activities in areas for which decontamination is most urgently required with respect to protection of human health and to implement such measures taking into account the existing levels of radiation. The Ministry of the Environment, as one of the implementing authorities, is coordinating and implementing remediation works giving due consideration to this policy on prioritization. However, the announcement made by the authorities shortly after the accident that "additional radiation dose levels should be reduced to annual exposure doses below 1 mSv in the long run" is often misinterpreted and misunderstood among people, both inside and outside the Fukushima prefecture. People generally expect that current additional radiation exposure doses should be reduced below 1 mSv per year immediately, as they believe that they are only safe when the additional dose they receive is below this value.

Finally, the team gave the following advice: Japanese institutions are encouraged to increase efforts to communicate that in remediation situations, any level of individual radiation exposure dose in the range of 1–20 mSv per year is acceptable and in line with international standards and with the recommendations from the relevant international organizations, such as ICRP, IAEA, UNSCEAR, and WHO. The government should strengthen its efforts to explain to the public that an additional individual exposure dose of 1 mSv per year is a long-term goal, and that it cannot be achieved in a short time, for example, solely by decontamination work.

There remains, however, worry about radiation exposure even after the IAEA advice. The authors think that the nonobjective explanation without understandable data clearly showing a safety of the dose "1 mSv/y" isn't effective in this case. Accordingly, it is worthwhile to publicize a communication to the people to clarify quantitatively the radiological safety of Nakadori district by evaluating the risk of "1 mSv/y" as well as natural radiation.

12.3.2 Community Dialog Forum for Residents of Fukushima Prefecture at Fukushima-shi [6]

This community dialogue forum was held on November 3, 2012 at Fukushima-shi for residents of the Fukushima prefecture to discuss matters such as returning their lives to normal in areas affected with long -term radiation from the TEPCO Fukushima Daiichi NPP accident, concentrating on the effects on health from exposure to radiation, with overseas experts mainly consisting of ICRP members who have expertise and knowledge in this field. After the key note lecture by the forum chair person of ICRP, several speakers from local media, the medical fraternity, and the residents belonging to various fields presented their actions just after the accident and problems they encountered. Almost all the speakers commonly talked about confusion due to lack of knowledge of radiation as well as poor information on the progress of the accident just after its occurrence. They felt strong anxiety regarding radiation. In the discussion, a few ICRP members mentioned that the radiation level of Fukushima was considerably lower in comparison with that of Chernobyl and they recommended comparing it with the natural radiation level. At that time, almost all Japanese had only limited information on natural radiation exposure doses such as the world average and higher values of India and China as well as the values in Japan. The higher values seemed to be not persuasive, because the Japanese knew neither such local high radiation areas in India and China nor the natural radiation levels in the European countries of the ICRP members. The material of Europe Atlas on natural radiation [10] which was informed after the community dialogue forum seemed to be doubtful to the authors, because the values given in the material were several times higher than the well-known world average value of 2.4 mSv/y. Accordingly, it is important to evaluate the country-averaged annual exposure dose by natural radiation for countries familiar to the Japanese. The country -averaged annual exposure dose is described in Sect. 12.5.

12.4 Risk Evaluation of 1 mSv/y for Public Radiation Exposure Limits

The latest recommendation for occupational and public radiation exposure limits have been made by ICRP in 2007 [11], taking account of the result of the long-term cohort study [12] on health effects in Japanese atomic bomb survivors in Hiroshima and Nagasaki, with a focus on not only "stochastic effects," primarily cancer, but also hereditary disorders. The data are given in three categories of exposure situations, namely, planned exposure situations that involve the deliberate introduction and operation of sources; emergency exposure situations that require urgent action in order to avoid or reduce undesirable consequences; and existing exposure situations that include prolonged exposure situations after emergencies. The most important quantities are the recommended dose limits in the planned exposure situations: occupational and public limits given in Table 12.3. The public exposure limit of "1 mSV/y" is taken as the lowest criterion to select the areas to be decontaminated and its risk is one of the most interesting matters to show safety.

The public limits were determined in order to secure excellent safety compared with those of other factors of mortality, on the basis of the concept proposed in 1977 [13]: (1) humans have always been exposed to radiation from the natural environment, the basic sources of natural radiation exposure. Man-made modifications of the environment and human activities can increase the "normal" exposure to natural radiation (2). Radiation risks are a very minor fraction of the total number of environmental hazards to which members of the public are exposed and the acceptable level of risk for stochastic phenomena for members of the general public may be inferred from consideration of risks that an individual can modify to only a small degree and which, like radiation safety, may be regulated by national ordinance. An example of such risks is that of using public transport. On this basis, a risk in the range of 10^{-6} to 10^{-5} per year would be likely to be acceptable to any individual member of the public.

In this chapter, the authors would certify the amount of radiation risk of public limits, "1 mSv/y" by using statistical data on Japanese mortality in 2008. The risk coefficient of the radiation dose of 100 mSv inducing cancer was estimated by using the death increase in a lifetime of 0.5 %, which was given by ICRP, the number of dead, 900,000, and the Japanese population, about 100 million. The results are compared with those of malignant tumors (cancer) and traffic accidents

Table 12.3 Recommended dose limits in planned exposure situations

Type of limit	Occupational, mSv in a year	Public, mSv in a year
Effective dose	20, averaged over 5 years, with no more than 50 mSv in any one year	1 (exceptionally, a higher value of effective dose could be allowed in a year provided that the average over 5 years does not exceed 1 mSv in a year)

Table 12.4 Risk coefficients of natural cancer, traffic accident and radiation exposure

Item	Mortality		Lifetime mortality (%)
	Number of deaths per 100,000 people	A fraction (Risk coefficient)	
Malignant tumors (Cancer)	270.1	2.7×10^{-3}	30.1
Traffic accident	5.9	5.9×10^{-5}	0.66
Radiation exposure of 100 mSv	4.5	4.5×10^{-5}	0.5

in Table 12.4. Because the limits of the occupational and public radiation exposure are one fifth and one hundredth of 100 mSv from the view point of total exposure dose, the risk coefficients were estimated as follows:

Occupational exposure limit 20 mSv/y: risk coefficient $= 9.0 \times 10^{-6}$,
Public exposure limit 1 mSv/y: risk coefficient $= 4.5 \times 10^{-7}$.

Consequently, it is noted that the risk coefficient of "1.0 mSv/y" is satisfactorily lower than 10^{-6} to 10^{-5} per year to be expected by ICRP.

12.5 Country-Averaged Annual Exposure Doses of Natural Radiation in the World

The ICRP members frequently said in the dialogue meetings with the residents of the Fukushima prefecture that the annual exposure doses in Fukushima were as low as those of natural radiation in their countries in Europe. The natural radiation doses in Europe are given in Reference [10] but they seem to be too large compared with the world -average value of 2.4 mSv/y reported in UNSCEAR 2000 [14]. Accordingly, the authors have evaluated the country-averaged annual exposure doses by the natural radiation in the world, on the basis of the fundamental data on indoor radon concentration, external exposure both outdoor and indoor, cosmic rays, and intake of food which were taken from Reference [14].

In the calculation, the authors assumed that a man stayed for 19.2 h (80 %) a day indoors and 4.8 h (20 %) outdoors, and that the concentration of outdoor radon was one third of that of indoor radon, as the conditions had been taken in the estimation of the world -averaged value. The conversion factor from the radon concentration to exposure dose was taken as $Q = 9.0 \times 10^{-6}$ (mSv/Bq m^{-3}) and a decay fraction of radon isotopes to the daughter nuclides contributing to actual radiation exposure as $k = 0.4$ that were ordinarily used in the estimation of the effect of natural radiation. Likewise, the conversion factor of a gamma-ray adsorbed dose to an effective equivalent dose was $C = 0.748$ (mSv/mGy). Annual exposure dose by cosmic rays was assumed to be 0.39 mSv/y for countries near the North Poleand

Table 12.5 Comparison of annual exposure dose due to natural radiation (unit: mSv/y)

Object		Radon	Cosmic rays	Indoor gamma	Outdoor gamma	Intake foods	Total
Japan	Published	0.48	0.30	—	0.33	0.98	2.09
	Present	0.53	0.30	0.07	0.28	0.98	2.16
World average	Published	1.26	0.39	—	0.48	0.29	2.42
	Present	1.17	0.38	0.08	0.44	0.29	2.35

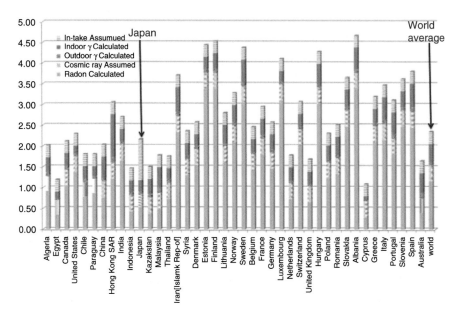

Fig. 12.2 Country averaged annual exposure dose rate due to natural radiation in the world

0.35 mSv/y, otherwise, except for 0.30 mSv/y for Japan. The annual dose due to food intake was also assumed to be the world -averaged value 0.29 mSv/y, except for 0.98 mSv/y for Japan: the Japanese eat many sea products, which contain radioactive isotopes such as polonium-210 and lead-210 and have high dose conversion factors in the human body.

In Table 12.5, the present results for Japan and the world average are compared with the widely published values. Good agreement is observed so that the authors might be convinced that the present method is verified. Figure 12.2 compares the calculated country-averaged annual exposure doses in the world. It is found that the values of northern and eastern European countries such as Finland, Estonia, Sweden, Luxemburg, Hungary, and Albania are quite high, exceeding 4 mSv/h because of high radon concentrations, whereas the values are lower than those of about 7 mSv/y given in Reference [10].

12.6 Discussion

The decontamination of a major part of living spaces of the "Intensive Contamination Survey Area" in the Fukushima prefecture has been done and ambient radiation dose rates have been measured to be below 0.23 μSv/h on average, whereas a higher radiation level is observed in a small part but does not exceed 0.5 μSv/h at the highest. In the special decontamination area of three municipalities such as Tamura-shi, Naraha-machi, and Kawauchi-mura, the average radiation dose rate was 0.38–0.41 μSv/h within FY 2014. At 15–30 % places, slightly high radiation has remained in the amount of 0.54–1.03 μSv/h measured immediately after the decontamination. The additional annual exposure dose for an individual due to radioactive cesium can be roughly estimated by multiplying 5000 h to the ambient radiation dose rate. The additional annual dose thus obtained can be expected to be below 1 mSv/y in most parts of the Nakadori district and 2 mSv/y on average even in the three municipalities.

The total annual dose that the resident will actually receive should be calculated by adding the contributions of natural radiation of about 2 mSv/y to the additional exposure dose due to radioactive cesium isotopes. It can be roughly said that the people in the Nakadori district will be exposed by 3 mSv per year maximum, and 4 mSv on average in the "Special Decontamination Area" of Tamura-shi, Kawauchi-mura, and Naraha-machi. A similar dose will be normally received in Europe, where radon concentration is high, as the ICRP members said.

The present work also clarified that the risk coefficient of 1 mSv/y is quite low, 4.5×10^{-7}, compared with that of traffic accidents of 5.9×10^{-5}. Even in the case of 7 mSv/y where the environmental radiation dose rate is about 1 μSv/h observed in Kawauchi-mura, its risk coefficient is 3.2×10^{-6} and seems to be low enough. Of course, the occupational exposure dose limits are also at a safe level: their risk coefficient 9.0×10^{-6} is one order lower than that of traffic accidents.

Accordingly, the authors hope the present results help people accept the reasonable decontamination work and not aim for instantaneously realizing 1 mSv/y and determine their return back to their hometowns in an environment of a few mSv/y.

Nevertheless, it is considered that the people who moved would be afraid of the influence of low-level radiation on cancer in their children and hesitate to return to their hometown. Low-level radiation brings a stochastic influence on health, essentially increasing death due to cancer with the probability of 0.5 % per 100 mSv in accordance with the linear model with the nonthreshold hypothesis proposed by ICRP [11]. On the other hand, the lifetime mortality of malignant tumors (cancer) is quite high because many stresses in daily life produce much reactive oxygen in a human body. The reactive oxygen damages the DNA of a cell in the human body; a double-strand break is especially a problem likely to produce a cancerous mother cell. Inasmuch as almost all DNA damage is repaired and imperfectly repaired DNA causes cell death (loss of cells) through apotosis, a person will seldom get the cancer, additively owing to his immunity to guard his body against impermissible different kinds of cells. He may get the cancer by losing immun ability as he ages. Laughter is helpful in strengthening immunity power by increasing killer cells in the human body. Accordingly, spending a daily life without so much stress is important for a person to protect against cancer by reducing reactive oxygen production and keeping immunity power. There is research that shows the stress of daily life produces a double-strand break of the DNA by about 300 times as much as 1 mSv/day radiation exposure does [15]. It might be evidence of the high lifetime mortality of malignant tumors, shown in Table 12.4. Anyhow, it can be said that the influence of low-level radiation below 100 mSv is not so large.

It is a difficult problem how we consider the influence of the total exposure dose of low-level radiation accumulated during a human lifetime. For example, the total dose by natural radiation of 2.1 mSv/y is estimated to be 150 mSv by taking into account agewise sensitivities to radiation. Its risk coefficient can be estimated to be 6.75×10^{-5} according to the LNT hypothesis which was proposed for radiation management in the radiation facilities by the ICRP, and the radiation effect is considered to increase proportionally to the radiation exposure dose. However, there is an idea that the influence of low-level radiation is not transmitted for a long time because its effects are eliminated for a short time by the human body guard system such as the functions of repairing the damaged DNA, immunity against the inimical cells, and so on. Recently, Banto al. [16] discussed human body recovery due to repairing damaged DNA and showed that the radiation effect becomes saturated to certain values. This idea implies that the present risk estimation for the public limits, 1 mSv/y, is always applicable to long-time exposure without integrating the risk per year.

12.7   Conclusion

In the present work, the additional radiation exposure dose in Fukushima is clarified. Newly evaluated country-averaged annual exposure dose by natural radiation and risk coefficient of 1 mSv/y also showed that the Fukushima Nakadori district is safe.

The decontamination work is being carried out throughout the living space of the Fukushima prefecture aiming at a goal in FY 2018. Recent measurements of the environmental radiation dose rates showed that the additional annual exposure dose was generally below 1 mSv/y in the major part of the Nakadori district and about 2.0 mSv/y in the special decontamination area of Tamura-shi, Naraka-machi, and Kawauchi-mura. The total exposure dose taking account of both cesium and natural radiation, 3–4 mSv/y are the same as the dose from natural radiation in Europe. There are no data to show any correlations between such country-averaged exposure doses by natural radiation and the cancer death data which can be taken from Reference [17]. It means that the low-level exposure of natural radiation does not cause any cancers.

The risk of the 1 mSv/y exposure has been certified to be quite low, 4.5×10^{-7} per year, as the ICRP 1977 recommendation expected as a condition of the public exposure limit.

This information should give a light to the residents to live in the Fukushima prefecture in the future. The authors would expect the people outside Fukushima, especially residents of the Tokyo metropolitan area, to correctly understand the risk coefficient of the 1 mSv/y and to recognize that the radiation level of Fukushima becomes lower to the harmless level, and finally to abandon their negative consciousness about Fukushima.

The major cause of cancer is stress in daily life such as an irregular life, unbalanced meals, friction with other people, smoking, and so on rather than radiation. A tranquil life with laughter is very important to protect from cancer rather than worrying about the influence of low-level radiation. Finally, the authors hope that the government and the media will accept the present result and publicize it widely in order to make both Fukushima and the whole of Japan brighter, through acceleration of the decontamination work and the rehabilitation together with the excellent ideas and passionate efforts to recreate a town in the damaged areas.

Acknowledgment The authors are highly grateful to Mr. Seiji Ozawa and Mr. Hiroyuki Suzuki MOE for their information about the progress of the decontamination work. They also thank Ms. Emi Konno and her assistant in the evaluation work on the natural radiation effect.

References

1. Fukushima Prefectural Office (2015) Present status of decontamination progress in area performed by the municipalities at the end of February 2015 (in Japanese)
2. Fukushima Office for Environmental Restoration of MOE Outline of progress of the national government implement measures in accordance with the Act on Special Measures concerning the Handling of Radioactive Pollution on January 1, 2012, 6th Materials distributed in the lecture meeting to pressmen on 14 April 2015. (in Japanese)
3. Recovery of Fukushima, Fukushima Prefectural Office (2015) Steps for revitalization in Fukushima, April 21, 2015 publication. http://www.pref.fukushima.lg.jp/uploaded/attachment/112884.pdf, (in Japanese)
4. Fukushima Prefectural Office (2015) Fukushima prefecture evacuee intention survey results (summary version), [PDF file /1, 285KB]. http://www.pref.fukushima.lg.jp/uploaded/attachment/113135.pdf, (in Japanese)
5. IAEA Mission Team (2014) The follow-up IAEA international mission on remediation of large contaminated areas off-site the Fukushima Daiichi Nuclear Power Plant, NE/NFW/2013
6. Kawai M (2012) Report on community dialog forum for residents of Fukushima prefecture with International experts on returning life to normal in areas affected with long term radiation from the Fukushima nuclear accident —the importance of the involvement of residents in returning lives and the environment to normal in areas with radioactive contamination —, held on 3 Nov. 2012, Fukushima-shi. $ATOMO\Sigma$, 55:166–171 (2013) (in Japanese). https://www.env.go.jp/jishin/rmp/attach/icrp-session121029_mat01.pdf
7. HP of NSR (2015) Airplane monitoring results for Fukushima and the neighboring prefectures, Feb. 13, 2015. http://radioactivity.nsr.go.jp/ja/contents/11000/10349/24/150213_9th_air.pdf

8. HP of Fukushima Prefectural Office (2015) Radiation monitoring information in Fukushima prefecture, https://www.pref.fukushima.lg.jp/sec/16025d/kukan-monitoring.html
9. Homepage of MOE (2015) Progress of decontamination in accordance with the implementation plan in special decontamination area. http://josen.env.go.jp/area/index.html, (in Japanese)
10. Green BMR et al (1992) Natural radiation ATLAS of Europe. Radiat Prot Dosim 45:491–493
11. Valentin J (ed.) (2007) The 2007 recommendations of the international commission on radiological protection, Annals ICRP Publication 103
12. Preston DL et al (2003) Studies of mortality of atomic bomb survivors. Report 13: solid cancer and non-cancer disease mortality 1950–1997. Radiat Res 160:381–407
13. ICRP (1977) Recommendations of the international commission on radiological protection, ICRP Publication 26. Ann. ICRP 1 (3)
14. UNSCEAR (2000) Sources and effects of ionizing radiation, UNSCEAR 2000 report, vol. 1. United Nations Publication, New York
15. Institute of Industrial Ecological Sciences University of Occupational and Environmental Health (2012) Introduction to radiology, Ugent guide for public on radiation exposure in responding to TEPCO Fukushima DAIICHI NPP accident. WWW.uoeh-u.ac.jp/kouza/hosyaeis/hibakuguide.pdf, p 81, (in Japanese)
16. Banto M et al (2015) LNT hypothesis is not realized. – repairing mechanism works in cell level under low-level radiation. ATOMOΣ 57:252–258 (in Japanese)
17. Curado MP et al (2007) Cancer incidence in five continents vol IX, The tables age-standardized and cumulative incidence rates (three-digit rubrics). International Agency for Research on Cancer http://www.iarc.fr/en/publications/pdfs-online/epi/sp160/CI5vol9-14.pdf

Chapter 13
Indoor Deposition of Radiocaesium in an Evacuation Area in Odaka District of Minami-Soma After the Fukushima Nuclear Accident

Hiroko Yoshida-Ohuchi, Takashi Kanagami, Yasushi Satoh, Masahiro Hosoda, Yutaka Naitoh, and Mizuki Kameyama

Abstract The indoor deposition of radiocaesium was investigated for 27 wooden houses in eight areas of Kanaya, Mimigai, Ootawa, Ooi, Kamiyama, Kamiura, Ebizawa, and Yoshina in Odaka district of Minami-Soma, Fukushima Prefecture from November 2013 to January 2015. Odaka district is within a 20 km radius of the Fukushima Daiichi nuclear power plant (FDNPP), which used to be designated as a restricted area and has been designated as an evacuation area. Dry smear test was performed over an area of 100 cm^2 on the surface of materials made of wood, glass, metal, and plastic in the rooms and the surface of wooden structure in the roof-space. Approximately 1000 smear samples were collected in total; 89% of the smear samples obtained in the rooms exceeded the detection limit (0.004 Bq/cm^2) and a maximum value was evaluated to be 1.54 Bq/cm^2; 77% of the smear samples taken from the wooden structure in the roof-space exceeded the detection limit and a maximum value was evaluated to be 1.14 Bq/cm^2. Area differences in surface contamination were observed. Assuming that two horizontal phases of the room have uniform surface contamination with the maximum median

H. Yoshida-Ohuchi (✉) • T. Kanagami
Tohoku University, Sendai, Miyagi, Japan
e-mail: hiroko@m.tohoku.ac.jp; four020@gmail.com

Y. Satoh
Advanced Industrial Science and Technology, Tsukuba, Ibaraki, Japan
e-mail: yss.sato@aist.go.jp

M. Hosoda
Hirosaki University, Hirosaki, Aomori, Japan
e-mail: m_hosoda@cc.hirosaki-u.ac.jp

Y. Naitoh • M. Kameyama
Japan Environment Research Co., Ltd., Sendai, Miyagi, Japan
e-mail: yutaka-naitou@jer.co.jp; mizuki-kameyama@jer.co.jp

© The Author(s) 2016
T. Takahashi (ed.), *Radiological Issues for Fukushima's Revitalized Future*,
DOI 10.1007/978-4-431-55848-4_13

radioactivity observed in Kamiura (0.1 Bq/cm^2) for 27 houses investigated, the ambient dose equivalent rate for ^{134}Cs and ^{137}Cs in November 2013 was calculated as approximately 0.002 μSv/h.

Keywords Odaka district of Minami-Soma • Evacuation area • Wooden houses • Indoor surface contamination • Fukushima nuclear accident • Smear method

13.1   Introduction

The Great East Japan Earthquake of magnitude 9.0 and the tsunami on March 11, 2011 resulted in major damage to the Fukushima Daiichi nuclear power plant (FDNPP). From March 12 onward, various incidents at multiple units occurred including hydrogen explosions, smoke, and a fire [1] and a large amount of radioactive material was released into the environment and moved as a radioactive plume with the wind [1, 2]. In the evening on March 12, a reading of 20 μSv/h was observed from a measurement made at the joint government building of the city of Minami-Soma, and it is believed that the plume was first blown south by a weak northerly wind and then diffused to the north by a strong southerly wind [1]. In the absence of precipitation, the dispersed pollution caused dry deposition during the radioactive plume pass through the area. In particular, low airtightness in Japanese wooden houses can be the cause of indoor dry deposition [3]. Dry deposition is important as well as wet deposition when assessing the consequences of nuclear accident in the context of risk assessment, as dry deposition is close to where residents live and doses from deposited long-lived contaminants are usually the major long-term hazards. It is necessary for residents to know the level of indoor deposition when they plan temporary access or return home. In this study, the indoor deposition of radiocaesium was investigated for 27 wooden houses in eight areas in Odaka district of Minami-Soma. Then, we assessed the influence of surface contamination on the ambient dose equivalents.

13.2   Methods

13.2.1   Locations of Houses Investigated

From November 2013 to January 2015, the indoor deposition of radiocaesium was investigated for 27 wooden houses in eight areas of Kanaya, Mimigai, Ootawa, Ooi, Kamiyama, Kamiura, Ebizawa, and Yoshina in Odaka district of Minami-Soma, Fukushima Prefecture, as shown in Fig. 13.1. Odaka district, located on the southern end of Minami-Soma, is within a 20 km radius of the FDNPP and all residents were evacuated immediately after the evacuation instruction was issued on March 12, 2011 [1]. On April 22, 2011, Odaka district was shifted to legally

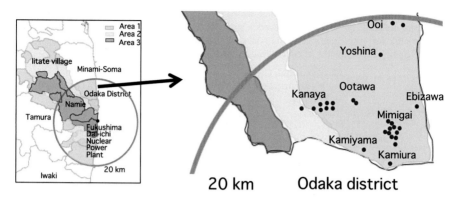

Fig. 13.1 Current status of Odaka district of Minami-Soma, Fukushima Prefecture in April 2015 and the locations of the houses. One *blue, closed circle* corresponds to one house at each location

enforceable restricted area [1]. On April 16, 2012, the restricted area was rearranged into three areas responding to the annual cumulative dose [1]. Areas 1, 2, and 3 are those in which the evacuation orders were ready to be lifted; in which the residents are not permitted to live; and where it is expected that the residents will have difficulty returning for a long time, respectively [4]. The Odaka residents were prohibited any access to their homes until April 2012. Decontamination work started from 2014 in Odaka district during the investigation of this study. It did not affect the investigation as it was conducted for the yard, the roof, and the gutter but not for the interior of the house. The current status of Odaka district on April 2015 and the locations of the houses investigated are shown in Fig. 13.1. One blue, closed circle corresponds to one house at each location. All houses are made of wood and are one- and/or two-story structures.

13.2.2   Measurement of Surface Contamination

To estimate surface contamination, dry smear test was performed on the surface of materials in the rooms and in the roof-space. The surfaces over an area of 100 cm^2 of materials were rubbed with moderate pressure using a round smear test paper with a diameter of 2.5 cm. The smear samples were carefully collected from the flat places that were not cleaned or wiped by residents, in every room, and 991 samples were collected in total. The numbers of houses and collected samples for surfaces of wooden, metal, glass, and plastic materials in the rooms, of wooden structure in the roof-space, and of wooden column in the rooms for each area are summarized in Table 13.1.

Radioactivity on smear test paper was measured for 5 min with a plastic scintillator detector JSD-5300 (Hitachi Aloka Medical, Ltd., Japan) and/or a liquid scintillation counter LS-6500 (Beckman Coulter, Inc.) using beta rays emitted from ^{134}Cs and ^{137}Cs, which are dominant nuclides in the investigated period.

Table 13.1 The numbers of houses and collected samples for surfaces of wooden, metal, glass, and plastic materials in the rooms, of wooden structure in the roof-space, and of wooden column in the rooms for each area and the numbers of smear samples exceeded the detection limit (0.004 Bq/cm^2) and below the detection limit for each area

	Numbers of houses	Wooden, metal, glass, and plastic materials in the rooms			Wooden structure in the roof-space			Wooden column in the rooms		
		Numbers of samples	<ND	>ND	Numbers of samples	<ND	>ND	Numbers of samples	<ND	>ND
Kamiura	1	31	2	29	2	0	2	2	2	0
Kamiyama	1	46	0	46	5	0	5	4	2	2
Kanaya	8	231	11	220	45	4	41	16	16	0
Ootawa	2	43	1	42	7	2	5	4	4	0
Yoshina	1	36	0	36	5	0	5	2	2	0
Ebizawa	1	29	4	25	1	0	1	2	1	1
Mimigai	11	334	47	287	52	22	30	18	14	4
Ooi	2	65	21	44	7	1	6	4	4	0
Total	27	815	86	729	124	29	95	52	45	7

In order to estimate the total removable surface contamination, radioactivity on smear test paper (Bq/cm^2) was obtained by the following equation [5, 6]:

$$A_{sr} = \frac{n - n_b}{60 \cdot \varepsilon_i \cdot F \cdot S \cdot \varepsilon_s} \qquad (13.1)$$

where n is the gross count rate (min^{-1}), n_b is the background count rate (min^{-1}), ε_i is the instrument efficiency, F is the removal fraction, S is the surface area covered by the smear, for example, 100 cm^2, and ε_s is the source efficiency, that is, the fraction of the decays within the sample that results in a particle of radiation leaving the surface of the source.

The combined efficiency of the instrument and the source for each detector of the plastic scintillator detector and the liquid scintillation counter was determined by measuring a part of samples simultaneously using a high-purity germanium (HPGe) detector ORTEC-GMX-20195-S (AMETEK Inc., USA). The removal fraction was determined by the use of repetitive wipes [5, 6] and the average $\pm\sigma$ of the values was evaluated to be 0.65 ± 0.28. All values of radioactivity were corrected to those in November 2013.

The detection limit, N_d, was defined as 3 standard deviation and calculated by the following equation [7]:

$$N_d = \frac{3}{2}\left[\frac{3}{T_s} + \sqrt{\left(\frac{3}{T_s}\right)^2 + 4N_b\left(\frac{1}{T_s} + \frac{1}{T_b}\right)} \right] \qquad (13.2)$$

where N_d is the background count rate (min^{-1}), T_s and T_b are the counting time of the sample and the background (min), respectively, for example, 5 min.

The lower detection limit for surface contamination was obtained when the smear samples were measured using the liquid scintillation counter. The detection limit was evaluated to be 0.004 Bq/cm^2 with Eqs. (13.1) and (13.2).

13.3 Results and Discussion

13.3.1 Indoor Surface Contamination for Odaka Houses

The numbers of smear samples, which were collected from surfaces of wooden, metal, glass, and plastic materials in the rooms, of wooden structure in the roof-space, and of wooden column in the rooms for each area, that exceeded the detection limit and below the detection limit in each area are summarized in Table 13.1. Eighty-nine percent (729/815) of the smear samples obtained in the rooms exceeded the detection limit and a maximum value was evaluated to be 1.54 Bq/cm^2. Seventy-seven percent (95/124) of the smear samples taken from wooden structure in the roof-space exceeded the detection limit and a maximum value was evaluated to be 1.14 Bq/cm^2. This result indicates that the pollution deposited not only in the interior of the house but also in the structure in the roof-space. However, only 13.5 % (7/52) of the smear samples exceeded the detection limit for wooden column in the rooms with a maximum value of 0.07 Bq/cm^2. This difference between former two results and the latter one indicates that radiocaesium deposition on a vertical surface is considerably lower than that on a horizontal surface. The median surface contamination with an interquartile range evaluated from surfaces of wooden, metal, glass, and plastic materials in the rooms for 27 houses in each area is shown in Fig. 13.2a. The interquartile range is expressed by Q1–Q3, which are the middle values in the first half and the second half of the rank-ordered data set, respectively. There seems to be a difference in the median surface contamination depending on areas. In the same manner, the median surface contamination with an interquartile range evaluated from surfaces of wooden structure in the roof-space for houses in each area is shown in Fig. 13.2b. The order of indoor surface contamination was the same between in the rooms and in the roof-space except a value for the house in Yoshina, as shown in Figs. 13.2a, b. Three areas of Ebizawa, Mimigai, and Ooi showing smaller values of the median surface contamination with an interquartile range in Fig. 13.2a, b are located closer to the ocean.

The houses of Kanaya and Mimigai showing a difference in the median surface contamination with an interquartile range in Fig. 13.2a, b are compared. The values of surface contamination in each house Kanaya and Mimigai, in which 8 and 11 houses were investigated, respectively, are shown in Fig. 13.3. The value of median surface contamination with an interquartile range for each house in two areas is exhibited in the left group and the right group, respectively. Generally, it was observed that the values of surface contamination in Kanaya tended to be

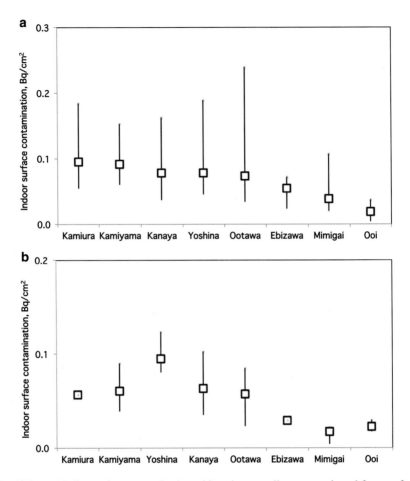

Fig. 13.2 (a) Median surface contamination with an interquartile range evaluated from surfaces of wooden, metal, glass, and plastic materials in the rooms for 27 houses in each area. The median and the interquartile range Q1–Q3 are expressed as *squares* and the *bar*, respectively. (b) Median surface contamination with an interquartile range evaluated from surfaces of wooden structure in the roof-space for houses in each area. The median and the interquartile range Q1–Q3 are expressed as *squares* and the *bar*, respectively

larger than those in Mimigai; however, a large discrepancy for each individual house was observed in both groups. The difference in characteristics of the houses, which showed the maximum and the minimum median surface contamination, in each group between Kanaya-A and Kanaya-B and between Mimigai-I and Mimigai-S was considered. There was no relationship between the value of surface contamination and the size of the house, the age of the building, the direction of the building (each entrance faces the south or the southeast direction), or area topography around the house. There is no difference in distance from the FDNPP among houses in each group, either. It was noted that residents of both, the houses

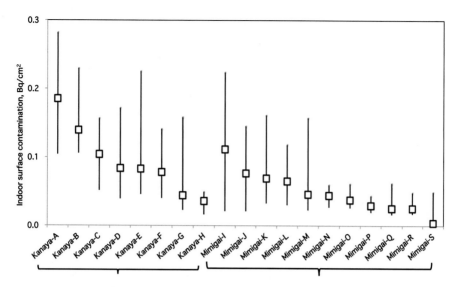

Fig. 13.3 Values of surface contamination in two groups of Kanaya and Mimigai. The value of median surface contamination with an interquartile range for each house in Kanaya and that in Mimigai is exhibited in the *left group* and the *right group*, respectively. The median and the interquartile range Q1–Q3 are expressed as *squares* and the *bar*, respectively

of Kanaya-B and Mimigai-S showing the minimum value of surface contamination in each group, could use running water and they quite often returned home after the restricted area was rearranged on April 2012. Even if the smear samples were carefully collected from places that were not cleaned or wiped, the frequency of residents' entering the house could have an effect on the level of surface contamination.

For houses in Kanaya and Mimigai, the ambient dose equivalents [$H^*(10)$] were measured outdoors using a $1''$ $\varphi \times 1''$ NaI (Tl) scintillation survey meter TCS-172B (Hitachi Aloka Medical, Ltd. Japan). $H^*(10)$ was measured outdoors at three or four points for each house and at a height of 1 m above the ground. At each point, measurements were collected by changing the direction of the probe of the survey meter to the four directions of east, west, north, and south, and each measurement was repeated three times. The measurement was conducted before the decontamination work started and all values were corrected to those in November 2013. An average $\pm\sigma$ was obtained for each house, as shown in Fig. 13.4, in the same order as the order of the data in Fig. 13.3. There is a clear discrepancy in the ambient dose equivalents between two groups of Kanaya and Mimigai, showing considerably lower values in the Mimigai group than those in the Kanaya group. A large discrepancy in the values of surface contamination for each house observed, seen in Fig. 13.3, is not seen in Fig. 13.4. It should be noted that Mimigai-I had a relatively low ambient dose equivalent, although the indoor surface contamination was relatively high.

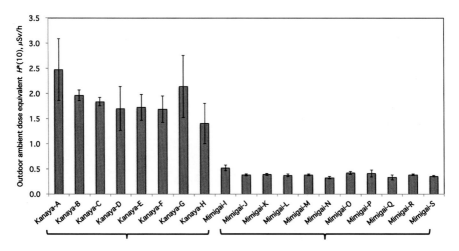

Fig. 13.4 Outdoor ambient dose equivalents [$H^*(10)$] for houses in Kanaya and Mimigai. An average $\pm\sigma$ for each house is shown in Fig. 13.4 in the same order as the order of the data in Fig. 13.3. The average and 1 standard deviation are expressed as *columns* and the *bar*, respectively

13.3.2   Effect of Surface Contamination on the Indoor Ambient Dose Equivalent

The effect of surface contamination on the indoor ambient dose equivalent was evaluated. The photon flux, φ (photons/sec/cm^2) at a point h m distant from the center of a disk-shaped isotropic source with a radius of R m, homogeneously emitting, is obtained by the following equation [8]:

$$\varphi = \frac{Q_s}{4} \ln \frac{R^2 + h^2}{h^2} \tag{13.3}$$

By applying 1 cm dose equivalent rate constant, Γ (μSv $\cdot$ m^2/MBq/h) to Eq. (13.3), the ambient dose equivalent rate, H (μSv/h) at a point h m distant from the center of a disk-shaped isotropic source with a radius of R m is obtained by Eq. (13.4), assuming radioactivity of 1.0 Bq/cm^2 is homogeneously distributed in the source,

$$H = \frac{\pi}{10^2} \Gamma \ln \frac{R^2 + h^2}{h^2} \tag{13.4}$$

Assuming two horizontal phases of the house with dimensions $10 \times 10 \times 3$ m are homogeneously contaminated with the maximum median radioactivity observed in Kamiura (0.1 Bq/cm^2) for 27 houses investigated, the ambient dose equivalent rate for ^{134}Cs and ^{137}Cs in November 2013 was calculated to be approximately 0.002 μSv/h.

The loose and removable surface contamination would cause not only external exposure but also internal exposure from intakes by ingestion and/or inhalation of radioisotopes. We evaluated an example of committed effective dose (μSv) using the radioactivity in dirt in a bag, which was sucked by a vacuum cleaner in the house (Kanaya-E) from the date before the earthquake occurred to November 2013 and effective dose coefficient given by ICRP 72 [9]. The radionuclide concentrations were counted using a high-purity germanium (HPGe) detector ORTEC-GMX-20195-S for 1000 s, revealing that concentrations for ^{134}Cs and ^{137}Cs were $68,600 \pm 0$ and $170,400 \pm 0$ Bq/kg (corrected to November 2013), respectively. In a case that an adult (more than 17 years) once intakes 0.1 g of dust by ingestion, the committed effective dose was calculated to be 0.35 μSv as ingestion dose coefficients are 0.019 and 0.013 μSv/Bq [9] for ^{134}Cs and ^{137}Cs, respectively.

Both, dose of external and internal exposure due to indoor deposition of radiocaesium are evaluated to be relatively low. Surface contamination is loose and easily removable. In our trials of cleaning, surface contamination for wooden materials was reduced to one-tenth after wiping with chemical wiping cloths and further reduced to below the detection limit after wiping with a damp cloth. From a viewpoint of radiation protection, it is strongly recommended to remove surface contamination by cleaning the house before residents return home.

13.4 Conclusion

The indoor deposition of radiocaesium was investigated for 27 wooden houses in eight areas in Odaka district using dry smear test. Comparison of surface contamination between Kanaya and Mimigai houses revealed that the values of surface contamination in Kanaya tended to be were larger than those in Mimigai; however, a large discrepancy for each individual house was observed in both groups. The frequency of residents' entering their house might have an effect on the level of surface contamination. We evaluated external and internal exposure dose due to indoor deposition of radiocaesium and found that these values are relatively low. Surface contamination is loose and easily removable. In our trials, surface contamination for wooden materials was reduced to below the detection limit after wiping with chemical wiping cloths and a damp cloth. From a viewpoint of radiation protection, it is strongly recommended to remove surface contamination by cleaning the house before the residents return home.

Acknowledgments This work was partly supported by a study of the Health Effects of Radiation organized by the Ministry of Environment, Japan. The authors thank Mr. Takemi Nemoto (Odaka ward office) for his help in recruiting house owners for investigation and Ms. Eri Hayasaka for her assistance in this study.

References

1. Nuclear Emergency Response Headquarters of Government of Japan (2011) Report of the Japanese Government to the IAEA ministerial conference on nuclear safety—the accident at TEPCO's Fukushima nuclear power stations. http://japan.kantei.go.jp/kan/topics/201106/iaea_houkokusho_e.html. Accessed 7 May 2015
2. Nuclear Emergency Response Headquarters of Government of Japan (2011) Additional report of the Japanese Government to the IAEA—the accident at TEPCO's Fukushima nuclear power stations—(Second report). https://www.iaea.org/sites/default/files/japanreport120911.pdf#search=\T1\textquoteleftNuclear+Emergency+Response+Headquarters+of+Government+of+Japan%2C+Additional+Report+of+the+Japanese+Government+to+the+IAEA\T1\textemdashThe+Accident+at+TEPCO\T1\textquoterights+Fukushima+Nuclear+Power+Stations\T1\textemdash%28Second+Report%29. Accessed 7 May 2015
3. Yoshida-Ohuchi H et al (2013) Evaluation of personal dose equivalent using optically stimulated luminescent dosemeters in Marumori after the Fukushima Nuclear Accident. Rad Protec Dosim 154:385. doi:10.1093/rpd/ncs245
4. Minister of Economy (2013) Trade and industry, evacuation areas, areas to which evacuation orders have been issued. http://www.meti.go.jp/english/earthquake/nuclear/roadmap/pdf/20130807_01.pdf. Accessed 7 May 2015
5. JIS Z 4504 (2008) Evaluation of surface contamination beta-emitters (maximum beta energy greater than 0.15 MeV) and alpha-emitters (in Japanese). http://kikakurui.com/z4/Z4504-2008-01.html. Accessed 7 May 2015
6. Frame PW, Abelquist EW (1999) Use of smears for assessing removable contamination. Health Phys 76(Supplement 2):S57–S66
7. Cooper JA (1970) Factors determining the ultimate detection sensitivity of Ge(Li) gamma-ray spectrometers. Nucl Instr Methods 82:273–277
8. Nakamura T (2001) Radiation physics and accelerator safety engineering, 2nd edn. Chijin-shokan, Tokyo (in Japanese)
9. ICRP (1995) Publication 72: Age-dependent doses to the members of the public from intake of radionuclides part 5, compilation of ingestion and inhalation coefficients: annals of the ICRP volume 26/1, 1e (International Commission on Radiological Protection)

Part IV
Radioactivity in Foods and Internal Exposure

Chapter 14
Radionuclides Behavior in Fruit Plants

Franca Carini, Massimo Brambilla, Nick G. Mitchell, and Hirofumi Tsukada

Abstract This paper summarizes research carried out on fruits by the Università Cattolica del Sacro Cuore (UCSC) in Piacenza, Italy. Among the fruit crops studied, strawberry, blackberry, grapevine, apple, pear, and olive, research on strawberry and blackberry was funded by the Food Standard Agency (UK). Fruit plants were grown in pots, kept under tunnels or in open field, and contaminated with ^{134}Cs and ^{85}Sr via leaves or via soil. Interception in strawberry plants ranges 39–17 % for ^{134}Cs, from anthesis (April) to predormancy (November). Leaf-to-fruit translocation occurs to a greater extent for ^{134}Cs than for ^{85}Sr. The distribution of contamination in fruit crops is an element-specific process: ^{134}Cs is preferentially allocated to fruits and ^{85}Sr to leaves. However, the activity in leaves is also species-specific: fruit species show different leaf-to-fruit translocation. Results on apple, pear, and grape crops indicate that the highest transfer from leaf to fruit occurs in apple crops. Olive plants also show ^{134}Cs translocation from leaves to trunks. Grapevines grown on mineral soil show a root uptake higher for ^{85}Sr than for ^{134}Cs, while strawberries grown on a peat substrate show a root uptake higher for ^{134}Cs than for ^{85}Sr. Rinsing directly contaminated fruits removes ^{85}Sr (36 %) to a greater degree than ^{134}Cs (24 %). Transfer to olive oil is low. A 57 % of ^{134}Cs is transferred from grapes to white wine.

F. Carini (✉)
Institute of Agricultural and Environmental Chemistry, Università Cattolica del Sacro Cuore,
Via Emilia Parmense 84, I-29122 Piacenza, Italy
e-mail: fcarini33@gmail.com

M. Brambilla
Institute of Agricultural and Environmental Chemistry, Università Cattolica del Sacro Cuore,
Via Emilia Parmense 84, I-29122 Piacenza, Italy

Consiglio per la ricerca in agricoltura e l'analisi dell'economia agraria (CRA), Unità di Ricerca per l'Ingegneria Agraria (CRA-ING), Laboratorio di Ricerca di Treviglio, Via Milano, 43, 24047 Treviglio, BG, Italy

N.G. Mitchell
Grant Harris Limited, PO Box 107, Haslemere, Surrey, GU27 9EF, UK

H. Tsukada
Institute of Environmental Radioactivity, Fukushima University, 1 Kanayagawa, Fukushima-shi, Fukushima 960-1296, Japan

© The Author(s) 2016
T. Takahashi (ed.), *Radiological Issues for Fukushima's Revitalized Future*,
DOI 10.1007/978-4-431-55848-4_14

Keywords [134]Cs • [85]Sr • Fruits • Interception • Leaf-to-fruit translocation •
Soil-to-fruit transfer • Wine-making • Food processing

14.1 Introduction

Radioactive contamination of the agricultural environment and the transfer of
radionuclides through the food chain, as a consequence of a nuclear release, may
require measures to protect human health and the environment. Understanding the
behaviour of radionuclides in these situations is crucial if these measures are to be
effective.

There are relatively few radioecological studies of fruit crops, which is surprising
given that they contribute about 8 % of world food production and have a high
economic value [1]. The fate of radionuclides in fruit systems is affected by a
combination of biological, chemical, and physical processes and is sensitive to the
timing of contamination relative to plant growth.

Contamination of fruits following an airborne release can occur from direct
deposition of radionuclides onto the fruit surface, from deposition onto other aboveground parts of the plant, or after deposition onto soil. The relative contribution
of these processes depends on many variables, such as the radionuclide involved,
the plant species, the plant phenological stage at time of deposition, and the soil
type. This topic was discussed under the IAEA BIOMASS program [2] and further
updated in IAEA TRS 472 [3] and the accompanying TECDOC-1616 [4].

Various projects on fruit production systems have been carried out by the
Università Cattolica del Sacro Cuore (UCSC) in Piacenza, Italy. Among the fruit
crops studied were: strawberry, blackberry, grapevine, apple, pear, and olive. These
studies considered plants grown in pots, kept under tunnels or in open field, and
contaminated with [134]Cs and [85]Sr via aboveground plant parts or via soil. Research
on strawberry and blackberry was funded by the Food Standard Agency (UK) and
carried out as a collaborative project between Mouchel (UK) and the UCSC [5]. The
objective of these projects was to understand the main processes that lead to fruit
contamination in the short term after deposition. Processes of interception, leaf-to-
fruit translocation and soil-to-fruit transfer are discussed in the following, as well
as processing activities associated with fruits after harvest, such as pressing olives
to produce oil and grapes for wine-making. Selected results from this research are
compared and summarized in this paper.

14.2 Interception

Results are presented on interception by strawberry plants, *Fragaria x ananassa*
Duchesne. Strawberry is a herbaceous perennial member of the rose family
(Rosaceae). The study used June bearer plants, cultivar Miss, which are

representative of commercial production in Italy. Young strawberry plants are transplanted at the end of July, develop flower buds in the autumn, lose all old leaves at the resumption of growth (in March), flower in the spring, and produce a single crop from early May to June. After harvest, plants are removed and replaced by new plants.

The aboveground part of different sets of strawberry plants was contaminated with ^{134}Cs and ^{85}Sr at various phenological stages: predormancy (November), preanthesis (early April), anthesis (end of April), and beginning of ripening (May). The fresh weight activity (Bq per kg fresh weight), biomass (g dry weight per plant), and leaf area (cm^2 per plant) were measured for each of the four replicates. Intercepted activity (%) was calculated from the ratio of activity deposited on the plant per total activity applied:

$$I\,(\%) = \frac{\text{Bq in plant component}}{\text{Bq applied} \cdot \text{plant}^{-1}} \times 100$$

Results for ^{134}Cs are reported in Fig. 14.1.

The results show that interception for ^{134}Cs is highest at anthesis (the period of flower development that starts with the opening of the flower buds), 39 %, and

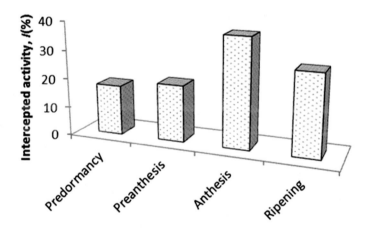

Phenological stage at the time of contamination

Fig. 14.1 ^{137}Cs intercepted by strawberry plants after deposition at different phenological stages: predormancy, preanthesis, anthesis, and ripening. $I(\%)$ is expressed as (Bq in plant component per Bq sprinkled $\cdot$ plant^{-1}) $\cdot$ 100

lowest at predormancy, 17 % [6]. Fruits, when present, have a lower interception capability than leaves, even if crop management, aimed at increasing the exposure of fruits to light, favors direct deposition and interception by fruit.

In a number of experimental studies, the interception of wet deposited radioactivity was found to be positively correlated with the leaf area and/or the dry biomass of the crop [7]. However, in the study described here there was no correlation between the interception capacity of strawberry plants and leaf area or biomass. Interception, at the growth stages considered, seems rather to be affected by variables such as the leaf senescence and the posture and physical orientation of leaves. A similar result was obtained contaminating tomato plants at two growing stages, demonstrating that an increase in the leaf surface area at later growing stages does not necessarily imply an increase in the level of interception of wet deposition [8].

Renaud and Gonze [9], studying the [134]Cs contamination of orchard fruits in Japan in 2011, observed that some correlation exists between cesium concentration in fruits, at first harvest, and ground surface deposit. They calculated the aggregated transfer factors, T_{ag}, expressed in $Bq \cdot kg^{-1}$ of fresh fruit per $Bq \cdot m^{-2}$ deposited on the ground surface. They observed that T_{ag} values of apricot samples were apparently higher when collected in low-elevation coastal areas (i.e., in Minamisoma-shi and Soma-shi municipalities, located to the North of the nuclear site, and Mito-shi to the South in Ibaraki Prefecture) than those collected from sites in mountainous areas. They suggested these differences were due to the stage of the vegetative cycle, as flowering would have occurred earlier in coastal regions than in inland elevated areas, resulting in greater transfer [9]. Similarly, the radiocesium activities of the foliar parts of woody species, 5 months after the Fukushima accident, were higher in evergreen species than in deciduous species, because the foliar parts of evergreen species (leaves from previous year) were present at the time of fallout but those of the deciduous species were not [10].

14.3 Leaf-to-Fruit Translocation

Fruits can receive radionuclides via translocation from contaminated aerial parts of plants, the most receptive of which, when present, is foliage. Leaf-to-fruit translocation has been studied in apple, pear, grapes, olives, blackberry, and strawberry plants, after contamination of leaves by wet deposition of [134]Cs and [85]Sr in open field conditions [6, 11–14]. The activity in fruits (Bq per kg fresh weight) was measured at ripening. Leaf-to-fruit translocation coefficients were calculated to represent the share of [134]Cs or [85]Sr activity found in fruits at ripening. Translocation coefficients, TC(%), are expressed as:

$$TC\,(\%) = \frac{Bq \cdot kg^{-1} \text{fresh weight fruit}}{Bq \text{ intercepted} \cdot plant^{-1}} \times 100$$

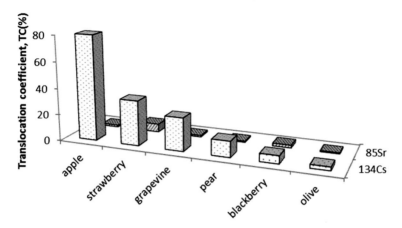

Fig. 14.2 Leaf-to-fruit translocation of ^{134}Cs and ^{85}Sr in various fruit crops. Translocation coefficients, TC(%), are expressed as (Bq kg^{-1} fresh weight fruit per Bq intercepted plant^{-1}) · 100

Results, reported in Fig. 14.2, indicate that the leaf-to-fruit TCs are on average one order of magnitude higher for ^{134}Cs than for ^{85}Sr in all fruit crops studied. This result is expected on the basis of greater foliar absorption of Cs^{+} than Sr^{2+} [15] and greater mobility, after absorption, in the phloem of monovalent ions than divalent ones [16, 17]. This makes the fruits sinks for cesium.

Differences in the process of foliar absorption and leaf-to-fruit translocation also exist between fruit species. Our results showed TCs for ^{134}Cs declined in the following order: apple trees, strawberry, grapevine, pear, blackberry, and olive plants (Fig. 14.2). A different trend was observed after the Chernobyl accident, in 1986, when Baldini et al. [18] found greater TCs for grapevine and peach than for apple and pear, and ascribed the differences to the more active metabolism of grapevine and peach.

However leaf-to-fruit translocation is also highly dependent on the time at which contamination occurs during the growth period of crops [19] and this can explain the results under different experimental conditions. In this regard, plants of strawberry and blackberry were contaminated with ^{134}Cs and ^{85}Sr via leaves at different growing stages: predormancy (November for strawberry and October for blackberry), anthesis (April for strawberry and May for blackberry), and beginning of ripening (May for strawberry and June for blackberry) [5, 6, 14, 20, 21]. Ripe fruits were picked and analyzed.

The growth cycle of strawberry plants was described earlier and a short description of blackberry plants is given in the following.

Blackberry is a bush plant, widespread in Northern Europe, with a perennial root apparatus and a biannual aerial part: the canes dry after having borne fruits. Blossoming is scalar, lasting up to 5–6 weeks. As a consequence, ripening is also very prolonged, from mid-July onward. The consumption of blackberry fruit is common particularly in northern countries, and can play a role in the diet of particular population groups.

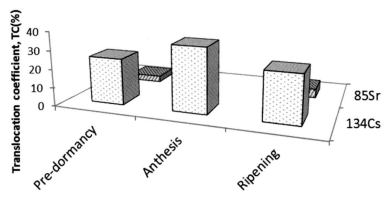

Fig. 14.3 Leaf-to-fruit translocation in strawberry: ^{134}Cs and ^{85}Sr activity translocated in fruit after leaf deposition at different phenological stages: predormancy, anthesis, and ripening. Translocation coefficients, TC(%), are expressed as (Bq kg^{-1} fresh weight fruit per Bq intercepted plant^{-1}) · 100

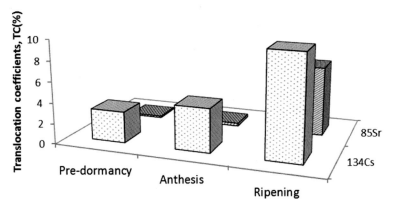

Fig. 14.4 Leaf-to-fruit translocation in blackberry: ^{134}Cs and ^{85}Sr activity translocated in fruit after leaf deposition at different phenological stages: predormancy, anthesis, and ripening. Translocation coefficients, TC(%), are expressed as (Bq kg^{-1} fresh weight fruit per Bq intercepted plant^{-1}) · 100

The activity (Bq per kg fresh weight) of fruits was determined and TCs(%) calculated; results are shownin Figs. 14.3 and 14.4 for strawberry and blackberry, respectively. The objective of this study was to show the dependence of plant contamination on the time of the year when the deposition occurs, referred to as seasonality by Aarkrog [22]. In strawberry, leaf-to-fruit translocationis greatest at anthesis, following the pattern observed for interception (Figs. 14.1 and 14.3). This result supports the hypothesis that interception is also affected by leaf age and by the metabolic activity of the plant. Increasing leaf age seems to reduce the absorption of radionuclides [23, 24].

When deposition occurs at predormancy, strawberry plants present well-developed leaves to aerial deposition, even if some are aging at this stage of the growing cycle. The plants have also developed flower buds for the following spring. A proportion of the radionuclides intercepted by, and absorbed into the leaves, is remobilized and translocated to the storage organs (roots and crowns) before leaf drop. Remobilization of ^{134}Cs is presumably higher than that of ^{85}Sr, given its higher mobility in the phloem. A few months later, the activity in ripe fruits is one order of magnitude greater for ^{134}Cs than for ^{85}Sr, an indication of the greater translocation of ^{134}Cs than ^{85}Sr from roots and crowns to fruits (Fig. 14.3).

In blackberry plants, the leaf-to-fruit translocation is greatest when foliar contamination occurs at ripening, and is one order of magnitude higher for ^{134}Cs than for ^{85}Sr (Fig. 14.4). When blackberry plant contamination occurs in Autumn, at predormancy, the plant system loses 60 % of the intercepted activity through dead leaves during Winter. A remaining 20 % is released into the soil and the environment. Fruit activity at harvest will be the result of retranslocation from roots and shoots (the storage organs) toward the fruits.

Recent observations after the Fukushima accident provide evidence that translocation to fruit does occur from leaf and bark, the amount of translocation differs between plant types, and that translocation is very sensitive to plant growth stage at the time of deposition [9, 25–27]. Another process that can contribute to fruit activity, ascertained by various authors after the Fukushima accident, is the secondary contamination, resulting from weathering, by the action of wind and rain [10, 27–30 cited by 28].

14.4   Soil-to-Fruit Transfer

Soil-to-fruit transfer was studied for grapevines, blackberry, and strawberry plants grown in pots [5, 11, 14, 21]. The crops were grown under different conditions using soils most suited to their cultivation. Data are not therefore comparable and are discussed separately as follows.

Root uptake for blackberry and strawberry plants was studied at two phenological stages: predormancy and anthesis. Results are presented as transfer factors, expressed as in IAEA [4]:

$$F_v = \frac{Bq \cdot kg^{-1} \text{ fresh weight fruit}}{Bq \cdot kg^{-1} \text{ dry soil}}$$

14.4.1 Grapevines

Two-year-old grapevines, variety Pinot Blanc, were grown in pots of 10 L capacity and kept in open field conditions. Pots were filled with a substrate of mineral soil (70 %) and sand (30 %). The mineral soil is moderately acidic (pH in H_2O 5.7), with low organic matter content (OM = 1.6 %) and low cation exchange capacity (CEC = 13.2 $cmol_{(+)}$/kg). The substrate of each pot was contaminated by moistening the surface with 250 mL of an aqueous solution containing 5305 kBq of ^{134}Cs and 2063 kBq of ^{85}Sr per pot after the fruit setting, at the beginning of July.

The objective of this study was to assess the transfer of ^{134}Cs and ^{85}Sr to fruit and other parts of the vine. At ripening, 60 days after soil contamination, F_v from soil to the whole plant was greater for ^{85}Sr than for ^{134}Cs, but while ^{85}Sr concentrates mainly in leaves and shoots, ^{134}Cs is redistributed throughout the plant. As a result, ^{134}Cs and ^{85}Sr F_v (mean ± standard error) in grapes are similar: $(8.0 \pm 0.9) \cdot 10^{-2}$ and $(6.6 \pm 1.2) \cdot 10^{-2}$, respectively [12].

14.4.2 Blackberry Plants

Two–year-old blackberry plants, cultivar Chester Thornless, were grown in 20 L capacity pots, filled with a mixture of peat (55 % of the total substrate), pumice, and compost. This is a medium with pH in H_2O 6.6, rich in organic matter (39.1 %), and with a high CEC (33.2 $cmol_{(+)}$/kg). Different sets of plants were contaminated in Autumn, at predormancy (October), and in Spring, at the beginning of anthesis (May). The soil surface was moistened with 800 mL of an aqueous solution containing 298 kBq of ^{134}Cs and 1921 kBq of ^{85}Sr per pot. The experiment lasted for 2 years following the growth cycle of this crop. The radionuclide content of fruit was determined at ripening, 276 or 74 days, respectively, after soil contamination.

The F_v are reported in Fig. 14.5. The F_v for ^{85}Sr is one order of magnitude greater than for ^{134}Cs and reflects the behavior of these radionuclides in mineral rather than organic soils [31]. The ratio of ^{85}Sr to ^{134}Cs in fruit was 12:1.

As for the phenological stages, a significant difference is shown ($p < 0.05$) for ^{85}Sr F_v mean and standard error: $(7.9 \pm 0.2) \cdot 10^{-1}$ at predormancy and $(5.9 \pm 0.2) \cdot 10^{-1}$ at anthesis; no significant difference is shown for ^{134}Cs: $(7.7 \pm 0.2) \cdot 10^{-2}$, at both plant stages.

14.4.3 Strawberry

Strawberry plants,cultivar MISS, were grown in 4.5 L capacity pots filled with peat, the normal substrate under horticultural growing conditions. Peat characteristics

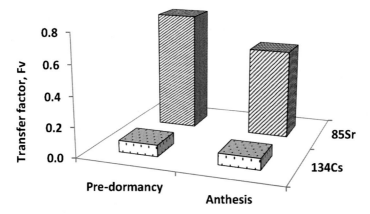

Fig. 14.5 Soil-to-fruit transfer factors F_v of ^{134}Cs and ^{85}Sr to blackberry following soil contamination at two phenological stages: predormancy and anthesis. F_v are expressed as (Bq kg^{-1} fresh weight fruit per Bq kg^{-1} dry soil)

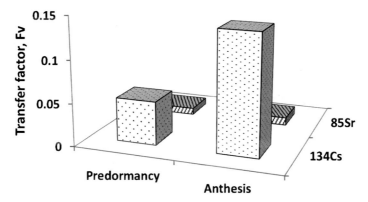

Fig. 14.6 Soil-to-fruit transfer factors F_v of ^{134}Cs and ^{85}Sr to strawberry following soil contamination at two phenological stages: predormancy and anthesis. F_v are expressed as (Bq kg^{-1} fresh weight fruit per Bq kg^{-1} dry soil)

were: a moderate acid pH (5.7), high OM (48.1 %), and high CEC (111.5 cmol$_{(+)}$/kg). The soil surface of each pot was moistened with 250 mL of an aqueous solution containing 74.8 kBq of ^{134}Cs and 91.6 kBq of ^{85}Sr at two phenological stages: predormancy (November) and anthesis (April). The experiment lasted for 3 years.

The radionuclide content of fruit was determined at ripening, 200 and 70 days respectively, after soil contamination. F_v are reported in Fig. 14.6 and show greater root uptake for ^{134}Cs, $(5.1 \pm 0.3) \cdot 10^{-2}$ at predormancy and $(1.4 \pm 0.1) \cdot 10^{-1}$ at anthesis, than for ^{85}Sr, $(7.8 \pm 0.9) \cdot 10^{-3}$ and $(9.2 \pm 1.6) \cdot 10^{-3}$, respectively. The ratio of ^{134}Cs to ^{85}Sr in fruit was 15:1.

The high organic matter content of the peat substrate, responsible for a large part of the CEC, reduces ^{134}Cs fixation on clay minerals, leaving it more available for root uptake, as highlighted by Nisbet and Shaw [31]. In contrast, it has been demonstrated that the organic matter of peat may form complex compounds with ^{85}Sr, reducing its availability to plants.

Differences in transfer to fruit after soil contamination at different phenological stages is apparent only for ^{134}Cs: F_v, are three times higher after contamination at anthesis than at predormancy (Fig. 14.6), the opposite was observed for blackberry fruits (Fig. 14.5).

Results on soil-to-plant transfer for blackberry and strawberry plants highlight time-dependent changes of root uptake following acute soil deposition during plant growth, as discussed by Choi [32]. These changes are more significant for an annual plant like strawberry, than for a perennial plant (with a perennial root apparatus) like blackberry and are more apparent for those radionuclides with the greatest transfer: ^{134}Cs on peat and ^{85}Sr on mineral soil.

14.5   Food Processing

Food processing losses can vary considerably depending on the type of radionuclide and between processes at the industrial and domestic scales. Rinsing directly contaminated apples (Golden Delicious) and grapes (Chardonnay) with tap water, simulating common domestic practice before consumption, removes ^{85}Sr to a greater degree than ^{134}Cs [13]. The processing factor P_f [3], called food processing retention factor, f_{fp}, in ICRU 65 [33], is the ratio of the radionuclide activity concentrations in the food after and before processing:

$$Pf = \frac{\mathrm{Bq \cdot kg^{-1}\ processed\ food}}{\mathrm{Bq \cdot kg^{-1}\ raw\ food}}$$

In our observations, it corresponds to 0.7 for ^{85}Sr and 0.8 for ^{134}Cs.

14.5.1   Wine-Making

A study of Pinot Blanc grapes, contaminated following an application of ^{134}Cs to leaves at the beginning of ripening, replicated the wine-making process on a small scale under laboratory conditions. Grapes were picked at the ripening stage, 35 days after contamination, analyzed for ^{134}Cs and squeezed to separate skin, pulp, and seeds from must (the juice extracted from grapes). The processing factor, P_f, for must was 0.94.

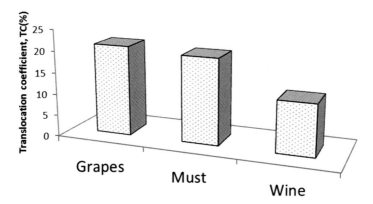

Fig. 14.7 Translocation coefficients, TC(%), from leaf to grapes, must, and wine for ^{134}Cs, expressed as (Bq kg^{-1} fresh weight per Bq intercepted plant^{-1}) · 100

Must was then inoculated (under controlled conditions) with 2.8 · 10^5 cells of the yeast Saccaromyces cerevisiae · mL^{-1} to begin the fermentation process. The must was fermented at 25 °C until constant weight was achieved. During the fermentation process, sugar changes into alcohol releasing CO_2 with a loss of weight:

$$C_6H_{12}O_6 \rightarrow 2\,(CH_3CH_2OH) + 2(CO_2) + \text{energy}$$

At the end of the process, the fermented must was filtered to remove any sediment. The ^{134}Cs content of wine corresponded to 61 % of the must activity (Fig. 14.7). Several different wine-making processes are practiced in industry, but all entail filtration, which is also carried out for home-made wine.

In practical situations the activity in wine can be quantified by the food processing factor P_f, if the activity of grapes is known, as follows: $P_f = (\text{Bq} \cdot \text{kg}^{-1}$ wine) per (Bq · kg^{-1} grapes). In this experimental study P_f for ^{134}Cs from grapes to wine was of 0.57. From the recent literature, the transfer of stable ^{133}Cs from rice to Japanese sake has been reported by Okuda et al. [34]. The authors calculated a P_f value of 0.04 from brown rice grains to sake.

14.5.2 Olive-Oil-Making

A second study was carried out using olive plants, *Gentile di Chieti*, contaminated via leaves with ^{134}Cs and ^{85}Sr at the beginning of ripening (at the end of September) [35]. At ripening, 10 days after contamination, olives were picked, analyzed for ^{134}Cs and ^{85}Sr, crushed by a mill, and pressed using a hydraulic jack, producing a fluid comprising oil, wastewater, and solid impurities. The oil was then separated from wastewater and solids by centrifugation. TCs were calculated for leaf-to-olives, leaf-to-wastewater, and leaf-to-oil. Results are shown in Fig. 14.8.

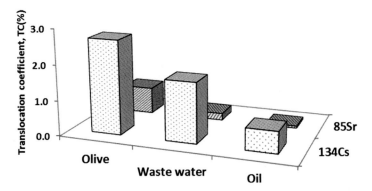

Fig. 14.8 Translocation coefficients, TC(%), from leaf to olive, wastewater, and oil, expressed as $(Bq \cdot kg^{-1}$ fresh weight per Bq intercepted $plant^{-1}) \cdot 100$

Only a small fraction of ^{134}Cs in fruit is transferred to oil. ^{85}Sr transfer to oil is not significant. P_f in this research, expressed as $(Bq \cdot kg^{-1}$ oil) per $(Bq \cdot kg^{-1}$ olives), gives values of 0.23 for ^{134}Cs and of 0.08 for ^{85}Sr, indicating that 77 % of ^{134}Cs and 92 % for ^{85}Sr are removed in the process of oil-making. Other studies on olive plants contaminated via soil report that a significant soil-to-fruit transfer of ^{134}Cs may occur, but no transfer to oil is detected [36]. Results from Cancio et al. [37] show that processing into olive oil removes ~90 % of the ^{134}Cs contamination initially contained in olive fruit. The P_f for olive oil obtained from data reported in the IAEA TRS 472 gives a value of 0.65 for ^{134}Cs [3].

14.6 Conclusions

From research at UCSC carried out on various fruit plants concerning the behavior of ^{134}Cs and ^{85}Sr after aerial or soil contamination, the following conclusions can be drawn:

- Interception seems to be affected by variables such as leaf senescence and the posture and physical orientation of leaves, rather than by leaf area or biomass.
- Leaf-to-fruit translocation is always higher for ^{134}Cs than for ^{85}Sr . It is also affected by the metabolic activity of the plant and by the phenological stage at time of contamination. The reproductive stages, like anthesis, lead to greater contamination of fruit.
- Soil-to-fruit transfer depends on the growing substrate. It is greater for ^{85}Sr on mineral soils, and greater for ^{134}Cs on peaty soils. Time-dependent changes of soil-to-fruit transfer following acute deposition are not insignificant in annual plants.
- The process of wine-making removes more than 40 % of ^{134}Cs present in grapes; that of oil-making removes about 75 % of ^{134}Cs and 90 % of ^{85}Sr in olives.

References

1. Salunkhe DK, Deshpande SS (1988) In: Salunkhe DK, Deshpande SS (eds) Foods of plant origin: production, technology and human nutrition. Van Nostrand Reinhold, New York, pp 1–5
2. International Atomic Energy Agency (2003) Modelling the transfer of radionuclides to fruit. Report of the fruits working group of BIOMASS theme 3, BIOsphere Modelling and ASSessment Programme, IAEA-BIOMASS-5, Vienna
3. International Atomic Energy Agency (2010) Handbook of parameter values for the prediction of radionuclide transfer in terrestrial and freshwater environments. Technical reports series no. 472. IAEA-TRS 472, Vienna
4. International Atomic Energy Agency (2009) Quantification of radionuclide transfer in terrestrial and freshwater environments for radiological assessments. IAEA-TECDOC-1616, Vienna
5. Food Standards Agency (2002) Experimental and modelling study on strawberry and blackberry, Draft final report, Mouchel Consulting Limited, Ref. 48100, Surrey, p 114
6. Carini F, Brambilla M, Ould-Dada Z, Mitchell NG (2003) 134Cs and 85Sr in strawberry plants following wet aerial deposition. J Environ Qual 32:2254–2264
7. Pröhl G, Hoffman FO (1996) Radionuclide interception and loss processes in vegetation. In: Modelling of radionuclide interception and loss processes in vegetation and of transfer in semi-natural ecosystems. Second report of the VAMP terrestrial working group, IAEA-TECDOC-857, Vienna, pp 9–48
8. Brambilla M, Fortunati P, Carini F (2002) Foliar and root uptake of 134Cs, 85Sr and 65Zn in processing tomato plants (Lycopersicon esculentum Mill.). J Environ Radioact 60:351–363
9. Renaud P, Gonze M-A (2014) Lessons from the Fukushima and Chernobyl accidents concerning the 137Cs contamination of orchard fresh fruits. Radioprot 49(3):169–175
10. Yoshihara T, Matsumura H, Hashida S-N, Nagaoka T (2013) Radiocesium contaminations of 20 wood species and the corresponding gamma-ray dose rates around the canopies at 5 months after the Fukushima nuclear power plant accident. J Environ Radioact 115:60–68
11. Carini F, Anguissola Scotti I, Montruccoli M, Silva S (1996) 134Cs foliar contamination of vine: translocation to grapes and transfer to wine. In: Gerzabek MH (ed) International symposium of the Austrian Soil Science Society: ten years terrestrial radioecological research following the Chernobyl accident, Vienna, 22–24 aprile, pp 163–169
12. Carini F, Lombi E (1997) Foliar and soil uptake of 134Cs and 85Sr by grape vines. Sci Total Environ 207:157–164
13. Carini F, Anguissola Scotti I, D'Alessandro PG (1999) 134Cs and 85Sr in fruit plants following wet aerial deposition. Health Phys 77(5):520–529
14. Fortunati P, Brambilla M, Carini F (2002) 134Cs and 85Sr in blackberry plants. Radioprotection – Colloques, Numéro spécial. In: Bréchignac F (ed) Proceedings of the international congress "The Radioecology – Ecotoxicology of Continental and Estuarine Environments, ECORAD 2001", Aix-en-Provence, 3–7 Sept 2001, vol 37, C1-547–552
15. Swietlik D, Faust M (1984) Foliar nutrition of fruit crops. Hortic Rev 6:299–301
16. Russel RS (1965) An introductory review: interception and retention of airborne material on plants. Health Phys 11:1305–1315
17. Marschner H (2002) Mineral nutrition in higher plants, 2nd edn. Academic, London
18. Baldini E, Bettoli MG, Tubertini O (1987) Effects of the Chernobyl pollution on some fruit trees. Adv Hortic Sci 1(2):77–79

19. Simmonds JR (1985) The influence of season of the year on the transfer of radionuclides to terrestrial foods following an accidental release to atmosphere. National Radiological Protection Board, NRPB-M121, Chilton Didcot

20. Fortunati P, Brambilla M, Carini F (2003) Trasferimento suolo-pianta di [134]Cs e [85]Sr in mora (Rubus fruticosus). Atti del Convegno annuale La conservazione della risorsa suolo. Piacenza, 8–10 giugno 2002. Bollettino della Società Italiana della Scienza del Suolo 52(1–2):363–372

21. Fortunati P, Brambilla M, Carini F (2003b) [134]Cs and [85]Sr in blackberry plants. In: Proceedings of the 7th international conference on the biogeochemistry of trace elements, 15–19 Jun 2003, vol 2. Slu, Uppsala, pp 376–377

22. Aarkrog A (1992) Seasonality. In: Modelling of resuspension, seasonality and losses during food processing. First report of the VAMP terrestrial working group, International Atomic Energy Agency, IAEA-TECDOC-647, Vienna, pp 61-96

23. Aarkrog A (1969) On the direct contamination of rye, barley, wheat and oats with 85Sr, 134Cs, 54Mn and 141Ce. Radiat Bot 9:357–366

24. Fortunati P, Brambilla M, Speroni F, Carini F (2004) Foliar uptake of [134]Cs and [85]Sr in strawberry as function by leaf age. J Environ Radioact 71/2:187–199

25. Nihei N (2013) Radioactivity in agricultural products in Fukushima. Chapter 8. In: Nakanishi TM, Tanoi K (eds) Agricultural implications of the Fukushima nuclear accident. Graduate School of Agricultural and Life Sciences, The University of Tokyo. Springer Open, Tokyo, pp 73–85. doi:10.1007/978-4-431-54328-2_14

26. Tagami K, Uchida S (2015) Effective half-lives of 137Cs from persimmon tree tissue parts in Japan after Fukushima Dai-ichi nuclear power plant accident. J Environ Radioact 141:8–13

27. Takata D (2013) Distribution of radiocesium from the radioactive fallout in fruit trees. Chapter 14. In: Nakanishi TM, Tanoi K (eds) Agricultural implications of the Fukushima nuclear accident. Graduate School of Agricultural and Life Sciences, The University of Tokyo. Springer Open, Tokyo, pp 143–162. doi:10.1007/978-4-431-54328-2_14

28. Toshihiro Yoshihara, Shin-nosuke Hashida, Kazuhiro Abe, Hiroyuki Ajito (2014a) A time dependent behavior of radiocesium from the Fukushima-fallout in litterfalls of Japanese flowering cherry trees. J Environ Radioact 127:34–39. doi:10.1016/j.jenvrad.2013.09.007

29. Tagami K, Uchida S, Ishii N, Kagiya S (2012) Translocation of radiocesium from stems and leaves of plants and the effect on radiocesium concentrations in newly emerged plant tissues. J Environ Radioact 111:65–69

30. Nakanishi TM, Kobayashi NI, Tanoi K (2012) Radioactive cesium deposition on rice, wheat, peach tree and soil after nuclear accident in Fukushima. J Radioanal Nucl Chem. doi:10.1007/s10967-012-2154-7 (Published Online)

31. Nisbet AF, Shaw S (1994) Summary of a 5-year lysimeter study on the time-dependent transfer of 137Cs, 90Sr, 239,240Pu and 241Am to crops from three contrasting soil types: 1. Transfer to the edible portion. J Environ Radioact 23(1):1–17

32. Choi YH (2009) Root uptake following acute soil deposition during plant growth. In: Quantification of radionuclide transfer in terrestrial and freshwater environments for radiological assessments. IAEA-TECDOC-1616, Vienna, pp 253–258.

33. International Commission on Radiation Units and Measurements, ICRU Report 65, (2001) Quantities, units and terms in radioecology. J ICRU 1(2):15

34. Masaki Okuda, Midori Joyo, Masafumi Tokuoka, Tomokazu Hashiguchi, Nami Goto-Yamamoto, Hiroshi Yamaoka, Hitoshi Shimoi (2012) The transfer of stable 133Cs from rice to Japanese sake. J Biosci Bioeng 114(6):600–605

35. Aquilano C (2001) Traslocazione di [134]Cs e [85]Sr in piante di olivo e nell'olio. Tesi di laurea. Università Cattolica del Sacro Cuore, Piacenza

36. Skarlou V, Nobeli C, Anoussis J, Haidouti C, Papanicolaou E (1999) Transfer factors of 134Cs for olive and orange trees grown on different soils. J Environ Radioact 45(2):139–147

37. Cancio D, Maubert H, Rauret G, Colle C, Cawse PA, Grandison AS, Gutierrez P (1993) Transfer of accidentally released radionuclides in agricultural systems (TARRAS). In: CEC (ed) CEC Euratom radiation protection programme progress report 1990–1991 EUR 14927 DE/EN/FR, Luxembourg, pp 579–600

Chapter 15
Effect of Nitrogen Fertilization on Radiocesium Absorption in Soybean

Naoto Nihei, Atsushi Hirose, Mihoko Mori, Keitaro Tanoi, and Tomoko M. Nakanishi

Abstract Radioactive materials that were released during the nuclear accident contaminated the soil and agricultural products. It has become clear that potassium fertilization is effective for the reducing radiocesium concentrations in agricultural crops. However, apart from reports about potassium, few reports have examined how nitrogen (N), which has a large effect on crop growth, contributes to the radiocesium absorption. Focusing on this point, we studied the effect of nitrogen fertilizer on the radiocesium absorption in soybean seedlings. The concentration of radiocesium in the seed of soybean was higher in nitrogen-fertilized plants than in plants grown without fertilizer. The radiocesium concentration in the aboveground biomass increased as the amount of nitrogen fertilization increased. A comparison of the effects of the different forms of nitrogen treatment shows that the highest radiocesium concentration in the aboveground biomass occurred with ammonium sulfate (approximately 3.7 times the non-N), the next highest absorption occurred with ammonium nitrate (approximately 2.4 times the non-N treatment), followed by calcium nitrate (approximately 2.2 times the non-N treatment). Furthermore, the amount of radiocesium in soil extracts was highest with ammonium-nitrogen fertilization. Further study is required to clarify the factors that incur an increase in radiocesium concentration in response to nitrogen fertilization. Special care is required to start farming soybean on fallow fields evacuated after the accident or on fields where rice has been grown before, which tend to have higher available nitrogen than the regularly cultivated fields.

Keywords Radiocesium • Soybean • Nitrogen • Ammonium

N. Nihei (✉) • A. Hirose • M. Mori • K. Tanoi • T.M. Nakanishi
Graduate School of Agricultural and Life Science, The University of Tokyo, 1–1–1, Yayoi, Bunkyo-ku, Tokyo, Japan
e-mail: anaoto@mail.ecc.u-tokyo.ac.jp

T. Takahashi (ed.), *Radiological Issues for Fukushima's Revitalized Future*,
DOI 10.1007/978-4-431-55848-4_15

15.1 Introduction

The Great East Japan Earthquake occurred on March 11, 2011, and it was immediately followed by the accident at the Fukushima Daiichi Nuclear Power Plant, Tokyo Electric Power Company. Radiocesium, the dominant nuclide released during the accident, reached agricultural lands in Fukushima and its neighboring prefectures and contaminated the soil and agricultural products [1, 2]. To guarantee the safe consumption and handling of agricultural, livestock, forestry, and marine products, monitoring inspections were established [3]. According to these inspections [4–6] the ratio of samples exceeding the new standard value of radiocesium [7] (100 Bq kg^{-1}) were found to be 5.7 % for soybean, 2.6 % for rice, and 11 % for wheat in 2011; 2.6 % for soybean, 0.0007 % for rice, and 0 % for wheat in 2012; and 1.9 % for soybean, 0.0003 % for rice, and 0 % for wheat in 2013. The inspection results indicate that the ratio of soybean exceeding 100 Bq kg^{-1} was high compared with that for rice and wheat and that the tendency to decline was low compared with that for rice and wheat. To revitalize agriculture, Fukushima Prefecture has been promoting the decontamination of agricultural lands; consequently, it has implemented an increase in the exchangeable potassium content in soil used to grow rice to approximately 25 mg 100 g^{-1} (dry soil) or higher. This occurred because it recently became clear that potassium fertilization was effective for reducing the radiocesium concentration in agricultural crops [8, 9]. Nitrogen (N) has a large effect on crop growth, and some reports have suggested that it also promotes the radiocesium absorption [10, 11]. However, few studies have examined how nitrogen contributes to the radiocesium absorption in soybean apart from potassium.

For the recovery and revitalization of agricultural industries, the analysis of radiocesium absorption in soybeans is necessary. Focusing on this point, we studied the effect of nitrogen fertilizers on the radiocesium absorption in soybean seedlings.

15.2 Materials and Methods

We cultured soybean (*Glycine max*) in a greenhouse (experiment 1) and in a field (experiment 2). For experiment 1, nitrogen fertilizer in the form of ammonium sulfate was applied at two levels: 0.4 and 1.3 g per 1 L pot (hereafter low-N and high-N, respectively). The radiocesium activity of the soil which was taken in Fukushima in August 2012 was approximately 30 kBq kg^{-1}, exchangeable potassium was 13.4 mg 100 g^{-1} soil, and pH was 6.0. The plants were grown until maturity, and the seed was collected. For experiment 2, soybean was grown in Iitate Village, Fukushima Prefecture. The radiocesium of the field was approximately 13 kBq kg^{-1} (15-cm depth), exchangeable potassium was 15.8 mg 100 g^{-1}, and pH was 6.2. Nitrogen fertilizer in the form of ammonium nitrate was applied at three levels: 0, 50, and 100 kg ha^{-1} (hereafter non-N, low-N, and high-N, respectively). We sowed the seeds for experiment 2 on June 16, 2014, and collected the aboveground biomass on September 2, 2014. Next, we studied the effect of the different forms of nitrogen treatment on radiocesium absorption by soybean seedling (experiment 3). Nitrogen was applied as calcium nitrate, ammonium nitrate, and ammonium sulfate at three levels: 0, 0.01, and 0.05 g per treatment (hereafter non-N, low-N, and high-N, respectively). We cultured soybean in a vessel ($6.5 \times 6.5 \times 6.5$ cm) for 18 days in a biotron (28 °C, 16 h light). We collected the aboveground biomass. Moreover, we studied radiocesium activity in soil extracts following nitrogen application (experiment 4). Nitrogen was applied to the soil at 0.5 g kg^{-1} as ammonium sulfate. We collected the soil after 1, 5, and 15 days, extracted it with 1 M calcium chloride, and then measured the radiocesium activity. We used soil sourced from the same batch in Experiment 1, 3, and 4. The radiocesium activities of all samples were measured using a sodium iodide scintillation counter (Aloka AM-300). In experiment 3, potassium, calcium, and magnesium concentrations in the soybean seedling were measured after acid decomposition using inductively coupled plasma optical emission spectrometry (ICP-OES) (PerkinElmer, Optima 7300).

15.3 Results

Figure 15.1 shows the effect of nitrogen fertilizer on radiocesium absorption by soybean. The concentrations of radiocesium in the seed (experiment 1) or the aboveground biomass (experiment 2) were higher in the high-N treatments than in the non-N or low-N treatments. Table 15.1 shows the influence of the different forms of nitrogen treatment on radiocesium activity in the aboveground (experiment 3). The highest activity occurred with ammonium sulfate (approximately 3.7 times the non-N treatment), the next highest activity occurred with ammonium nitrate (approximately 2.4 times the non-N treatment), followed by calcium nitrate (approximately 2.2 times the non-N treatment). The concentrations of radiocesium in the

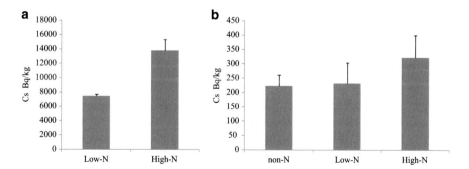

Fig. 15.1 (**a**) Experiment 1. The plants were cultured until maturity within 1 L pot, and the seed was analyzed. (**b**) Experiment 2. The plants were cultureed in Iitate village, Fukushima prefecture, and the aboveground biomass was analyzed

Table 15.1 Radiocesium activities and concentrations of base cations in soybean aboveground biomass after nitrogen fertilization treatments (Experiment 3)

		Cs	K	Ca	Na	P	Dry weight	Height
Treatment		Bq kg^{-1}	mg g^{-1}	mg g^{-1}	mg g^{-1}	mg g^{-1}	g plant^{-1}	cm
(Control)	Non-N	440	21	3.3	0.09	2.0	1.6	28
Calcium nitrate	Low-N	464	22	4.8**	0.09	2.0	1.6	30
	Hihg-N	984	24	8.1**	0.22*	1.5	1.3	17*
Ammonium nitrate	Low-N	642	22	3.6	0.12	2.6	1.4	25
	Hihg-N	1077**	19	3.4	0.09	2.3	1.6	29
Ammonium sulfate	Low-N	750	21	3.5	0.09	2.0	1.6	30
	Hihg-N	1634**	21	3.1	0.12	2.9	1.5	25

$^*p < 0.05$, $^{**}p < 0.01$ compared to the control (Dunnet's test)

Fig. 15.2 Radiocesium extracted from the soil after nitrogen fertilization (Experiment 4)

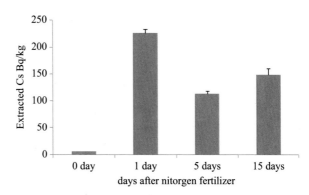

aboveground biomass were higher in nitrogen-fertilized plants than in plants without added nitrogen. Figure 15.2 shows the radiocesium activity in soil with applied nitrogen (experiment 4). The amount of extracted radiocesium increased a day after fertilization and remained higher even after 15 days of nitrogen application.

15.4   Discussion

The radiocesium concentration in seed and aboveground biomass increased as the amount of nitrogen fertilizer increased. The different forms of nitrogen treatment increased the radiocesium concentration of soybean in the order ammonium sulfate > ammonium nitrate > calcium nitrate. Hence, ammonium-nitorogen increased radiocesium absorption more than nitrate. Furthermore, the amount of radiocesium extracted from the soil, which is considered potentially available for plant absorption, was increased by ammonium-nitrogen fertilization.

Geometrically adapted cesium ions are fixed to the clay mineral, and the radiocesium adsorbed to the soil particles was probably not available for plant uptake. But the ionic radius of the ammonium ion is similar to that for the cesium ion [10], ammonium exchanged, and released radiocesium from the soil. We found that the amount of radiocesium extracted by ammonium fertilizer increased the day after fertilization; therefore, soybean could absorb radiocesium. In addition, the ammonium and cesium ions are both univalent cations, and ammonium has been found to restrict cesium absorption in the hydroponics [9–11]. This study was used by soil. It was considered that ammonium oxidizes to nitrate during cultivation, and ammonium fertilization did not restrict the radiocesium absorption of soybean. Moreover, we considered that potassium absorption might compete with ammonium absorption because both ions are univalent cations; hence, the soybeans may lack potassium. A lack of potassium has been found to increase cesium absorption [9, 12, 13]. However, the potassium concentrations in soybean did not decrease with ammonium fertilization. Therefore, we suggest that the increased activity of radiocesium in soybean due to ammonium fertilization was not because of a lack of potassium.

Soybean cultivation typically follows rice cultivation. To assist the recovery and revitalization of agricultural industries in Fukushima Prefecture, we suggest that special care is required to select the appropriate kind of fertilizer and to start cultivating soybean on fields that have higher available nitrogen than the regular cultivated fields. It is important to clarify the mechanism of cesium availability in response to nitrogen fertilizers to cope with cesium contamination in crops.

References

1. Yasunari TJ, Stohl A, Hayano RS, Burkhart JF, Eckhardt S, Yasunari T (2011) Cesium-137 deposition and contamination of Japanese soils due to the Fukushima nuclear accident. Proc Natl Acad Sci U S A 108:19530
2. Zheng J, Tagami K, Bu W, Uchida S, Watanabe Y, Kubota Y, Fuma S, Ihara S (2014) $^{135}Cs/^{137}Cs$ isotopic ratio as a new tracer of radiocesium released from the Fukushima nuclear accident. Environ Sci Technol 48:5433
3. Nihei N (2013) Radioactivity in agricultural products in Fukushima. In: Nakanishi TM, Tanoi K (eds) Agricultural implications of the Fukushima nuclear accident. Springer, Tokyo/New York, pp 73–85
4. Hamada N, Ogino H, Fujimichi Y (2012) Safety regulations of food and water implemented in the first year following the Fukushima nuclear accident. Radiat Res 53:641
5. Fukushima Prefecture (2011) (Toward a new future of Fukushima): http://www.new-fukushima.jp/monitoring/en/ Accessed 10 May 2015
6. Nihei N (2015) Monitoring inspection in Fukushima prefecture and radiocesium absorption of soybean (in Japanese). Isot News 73:118–123
7. Nihei N, Tanoi K, Nakanishi TM (2015) Inspections of radiocesium concentration levels in rice from Fukushima Prefecture after the Fukushima Dai-ichi Nuclear Power Plant accident. Scientific reports 5, Article number: 8653–8658(2015/3)

8. Zhu YG, Shaw G, Nisbet AF, Wilkins BT (2000) Effects of external potassium supply on compartmentation and flux characteristics of radiocaesium in intact spring wheat roots. Ann Bot 85:293–298

9. Sanchez AL, Wright SM, Smolders E, Nayor C, Stevens PA, Kennedy VH, Dodd BA, Singleton DL, Barnett CL (1999) High plant uptake of radiocesium from organic soils due to Cs mobility and low soil K content. Environ Sci Technol 33:2752–2757

10. Tensyo K, Yeh KL, Mitsui S (1961) The uptake of strontium and cesium by plants from soil with special reference to the unusual cesium uptake by lowland rice and its mechanism. Soil Plant Food 6:176

11. Ohmori Y et al (2014) Difference in cesium accumulation among rice cultivars grown in the paddy field in Fukushima Prefecture in 2011 and 2012. J Plant Res 127:57–63

12. Cline JF, Hungate FP (1960) Accumulation of potassium, cesium137, and rubidium86 in bean plants grown in nutrient solutions. Plant Physiol 35:826. doi:10.1104/pp.35.6.826

13. Evans EJ, Dekker AJ (1969) Effect of nitrogen on cesium 137 in soils and its uptake by oat plants. Can J Soil Sci 49:349–355

Chapter 16
Concentrations of $^{134,\ 137}$Cs and ^{90}Sr in Agricultural Products Collected in Fukushima Prefecture

Hirofumi Tsukada, Tomoyuki Takahashi, Satoshi Fukutani, Kenji Ohse, Kyo Kitayama, and Makoto Akashi

Abstract On April 1, 2012, new Standard Limits for radionuclide concentrations in food were promulgated, superseding the Provisional Regulation Values in Japan set in 2011. The new Standard Limits are calculated based on 1 mSv y^{-1} of annual internal radiation dose through food ingestion of ^{134}Cs, ^{137}Cs, ^{90}Sr, Pu and ^{106}Ru, which were detected or possibly released into the environment from the accident at the TEPCO Fukushima Daiichi Nuclear Power Stations (FDNPS). The concentrations of the radionuclides were based on the values of radiocesium ($^{134,\ 137}$Cs) and of the other radionuclides (^{90}Sr, Pu and ^{106}Ru); the ratio observed in the determination or predicted concentrations in the soils from the FDNPS accident was used for estimating the concentration of the other radionuclides by means of the ratio against ^{137}Cs. The new Standard Limit of radiocesium in general foods was defined to be 100 Bq kg^{-1} fresh weight by the Ministry of Health, Labour and Welfare. In the present study the concentration of radiocesium was measured in agricultural products collected mostly in Fukushima-shi and Date-shi, Fukushima Prefecture, in 2012 and 2013. The average concentration of radiocesium in agricultural plants in 2012 was 7.6 (<0.2–40) Bq kg^{-1} fresh weight, decreasing to 2.0 (<0.1–14) Bq kg^{-1} fresh weight in 2013, which was approximately one-fourth of the concentration in 2012. The concentration of ^{90}Sr in agricultural products

H. Tsukada (✉)
Institute of Environmental Radioactivity, Fukushima University, 1 Kanayagawa, Fukushima-shi, Fukushima 960-1296, Japan
e-mail: hirot@ipc.fukushima-u.ac.jp

T. Takahashi • S. Fukutani
Research Reactor Institute, Kyoto University, 2 Asashiro-Nishi, Kumatori-cho, Sennan-gun, Osaka 590-0494, Japan

K. Ohse • K. Kitayama
Fukushima Future Center for Regional Revitalization, Fukushima University, 1 Kanayagawa, Fukushima-shi, Fukushima 960-1296, Japan

M. Akashi
National Institute of Radiological Sciences, 4-9-1 Anagawa, Inage-ku, Chiba-shi, Chiba 263-8555, Japan

© The Author(s) 2016
T. Takahashi (ed.), *Radiological Issues for Fukushima's Revitalized Future*,
DOI 10.1007/978-4-431-55848-4_16

collected in Fukushima Prefecture in 2013 was 0.0047–0.31 Bq kg^{-1} fresh weight, which was a similar range to those collected throughout Japan. The concentration ratio of ^{90}Sr/^{137}Cs in the agricultural plants collected from the area 5 km west from the Nuclear Power Stations (difficult-to-return zone) was lower than the predicted ^{90}Sr/^{137}Cs ratio, which was calculated using the ratio in the soils and soil-to-plant transfer factors.

Keywords Agricultural product • New standard limits • $^{134,\,137}$Cs • ^{90}Sr • ^{90}Sr/^{137}Cs ratio

16.1 Introduction

Significant quantities of radionuclides were released into the environment from the TEPCO's Fukushima Daiichi Nuclear Power Stations (FDNPS) accident in March 2011. Radiocesium ($^{134,\,137}$Cs) is the major radionuclide released by the accident and an important radionuclide for the assessment of radiation exposure to the public. Other relatively long half-life radionuclides such as ^{90}Sr, Pu, etc. are also important radionuclides for radiation dose estimation through long-term food ingestion. The new Standard Limits for radionuclides in foods was established by the Ministry of Health, Labour and Welfare on April 1, 2012. The limits were determined on the basis of 1 mSv y^{-1}. The limit in general foods is 100 Bq kg^{-1} of radiocesium, including the contribution of ^{90}Sr, Pu and ^{106}Ru, determined by using the actual concentration in the soils from the accident (estimated data for ^{106}Ru) and soil-to-transfer factor. However, the public has been concerned about food contamination, especially ^{90}Sr. In the present study the concentrations of radiocesium and ^{90}Sr in agricultural and animal products produced in Fukushima Prefecture were determined, and compared with the values of the new allowable Standard Limits.

16.2 Materials and Methods

Agricultural and animal products, limited to those produced within the Fukushima Prefecture, were collected from markets located mostly in Fukushima-shi and Date-shi. Table 16.1 contains a list of 120–5000 g samples of 11 spices in 2012 (40 samples) and 2013 (42 samples). The agricultural plants were washed, peeled, and then the edible parts were cut into small pieces. Each sample was dried at 70 °C for 1 week and pulverized in a stainless steel cutter blender before being analyzed for radiocesium. The animal samples were also cut into small pieces. The samples were compressed into a plastic container (47 mm in diameter and 50 mm in height) and the concentration of radiocesium and ^{40}K determined with a Ge detector connected to a multichannel analyzer system by counting for 9400–33,000 s. The

Table 16.1 Collected agricultural and animal products in Fukushima Prefecture in 2012 and 2013

Agricultural and animal products	Collected sample
Rice	Brown rice
Potatoes	Potato, Sweet potato, Eddoe
Savory herbs	Japanese ginger
Leaf and stem vegetables	Komatsuna, Malabar spinach, String bean, Cabbage, Welsh onion, Leek, Spinach, Japanese honeywort, Turnip (leaf and stem), Turnip rape, Mugwort, Asparagus, Onion
Root vegetables	Turnip, Burdock, Radish, Carrot
Pulses	Green soybean, Green bean, Cowpea, Black soybean
Fruity vegetables	Cucumber, Tomato, Green pepper, Eggplant, Pumpkin, Small green pepper, Okra, Broccoli, Snap garden pea, Zucchini, Wax gourd
Fruits	Pear, Apple, Persimmon, Huckleberry, Plum, Peach, Blue berry, Japanese plum
Wild vegetables	Udo, Butterbur, Momijigasa
Other agricultural plants	Jew's-ear mushroom, Shiitake mushroom, Hen-of-the-woods mushroom, Edible chrysanthemum, Japanese pepper (leaf)
Animal products	Chicken, pork, egg

detection efficiency of the Ge detector was dependent on the sample thickness and was obtained using the mixed standard radionuclides material made by the Japan Radioisotope Association. Counting statistics standard deviations for ^{137}Cs concentration in the sample were less than 10 % of the value.

Soil and agricultural samples were collected from 10 agricultural fields located both outside and within the more than 50 mSv y^{-1} of the external radiation dose zone (difficult-to-return zone, Okuma) in Fukushima Prefecture in 2013 and 2014 (Fig. 16.1) . Shiitake mushroom and its mushroom bed for cultivation was also collected. A stainless steel core sampler was used to collect soil cores 5 cm in diameter and 20 cm in depth at 5 points evenly distributed in each field. Twenty kilogram of each agricultural sample was collected from each field. The soil core samples collected from each field were dried at 50 °C for 1 week and then passed through a 2 mm sieve. The soil samples in each field were thoroughly mixed. The agricultural samples were washed, peeled, and then the edible parts were cut into small pieces. Approximately 100 g of dried sample was pulverized in a stainless steel cutter blender before being analyzed for radiocesium. The rest of the dried agricultural samples were washed at a temperature below 450 °C for analysis of ^{90}Sr. The dried soil and agricultural samples were compressed into plastic containers and the concentrations of radiocesium and ^{40}K determined. A radioanalytical method for ^{90}Sr was performed according to the previously reported method [1]. The ash plant samples (20–50 g) were decomposed with HNO_3, H_2O_2 and HCl after the addition of the Sr carrier. The soil samples (100 g) were heated at 450 °C and then extracted with 12 M HCl after the addition of the Sr carrier. The solution was filtered and the residue discarded. The solution was adjusted to >pH 10 with

Air dose rate at 1 m above the ground (µSv h⁻¹, November 19, 2013) [5]

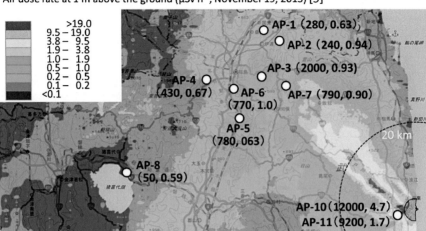

Fig. 16.1 Concentrations of ^{137}Cs and ^{90}Sr (Bq kg^{-1} dry weight) in cultivated soil ($n = 10$) collected from Fukushima Prefecture in 2014. Point numbers show the sampling agricultural plants indicated in Table 16.3 and the values in parentheses are the concentrations of ^{137}Cs (former) and ^{90}Sr (latter) in cultivated soil

NaOH and then SrCO$_3$ precipitated by adding Na$_2$CO$_3$. The SrCO$_3$ precipitate was dissolved in HCl and then the oxalates re-precipitated at pH 4.2 by adding oxalic acid. The supernatant was decanted and the oxalate precipitation dissolved in HNO$_3$. Strontium in the solution was separated from Ca by the cation ion-exchange method, and then the filtrated precipitate was dissolved with water. Any radioactive impurity was eliminated by scavenging on BaCrO$_4$ and Fe(OH)$_3$. An ammonium carbonate solution was added to the solution after scavenging, and the SrCO$_3$ precipitate was filtered using a cellulose filter paper. The recovery of Sr was estimated by measuring the stable Sr with ICP-AES and then a disk sample prepared for beta-counting for 6000–60,000 s.

16.3   Results and Discussion

The concentrations of radiocesium in agricultural products collected in Fukushima Prefecture in 2012 and 2013 are listed in Table 16.2. The average concentration of radiocesium collected in 2013 was 2.0 ± 2.7 (<0.046–14) Bq kg^{-1} fresh weight, which was one-fourth of that in 2012 (7.6 ± 10, <0.11–40 Bq kg^{-1} fresh weight), and the concentrations of radiocesium in all the samples were less than the new Standard Limits. The concentrations of radiocesium in agricultural products decrease with time elapsed, and the reasons are as follows:

Table 16.2 Concentration ranges of ^{134}Cs, ^{137}Cs and ^{40}K in agricultural and animal products (grouping noted in Table 16.1) collected from Fukushima Prefecture in 2012 and 2013

| Agricultural and animal products | 2012 | | | | 2013 | | | |
| | ^{134}Cs | ^{137}Cs | ^{40}K | n | ^{134}Cs | ^{137}Cs | ^{40}K | n |
	Concentration (Bq kg^{-1} fresh weight)				Concentration (Bq kg^{-1} fresh weight)			
Rice	1.0–2.5	1.4–4.9	34–55	3	<0.58	0.66	82	1
Potatoes	0.25–2.8	0.53–4.4	120–190	3	0.42–1.6	0.86–3.5	110–160	3
Savory herbs				–	1.2	2.4	130	1
Leaf and stem vegetables	0.08–2.0	0.17–3.8	39–140	8	<0.11–2.8	<0.11–4.7	42–280	15
Root vegetables				–	<0.022–0.41	<0.024–0.83	78–140	5
Pulses	5.9–15	10–25	160–560	3	1.1	2.2	210	1
Fruity vegetables	<0.063–2.3	<0.052–3.6	48–180	8	<0.046–0.85	<0.10–2.1	53–180	9
Fruits	0.16–13	0.25–23	26–170	7	0.77–0.92	1.6–2.1	45–47	2
Wild vegetables					0.26–4.4	0.51–9.6	94–160	4
Other agricultural plants	1.4–5.4	2.4–8.8	18–100	4	1.1	2.1	110	1
Animal products	<0.85	<0.68	60–110	4				–

1. Decay of radiocesium activities, especially ^{134}Cs (half-life, 2.1 y)
2. Countermeasure with the application of K fertilizer for reducing uptake of radiocesium in plants
3. Aging effect in soil, which radiocesium in exchangeable fraction decreases and that in strongly bound fraction increases with time elapsed [2, 3]
4. Decreasing radiocesium contents in soil by erosion, etc.

The concentration of radiocesium in agricultural plants drastically decreases immediately after the accident, and the rate of the decrease of radiocesium concentration in plants has gradually slowed as more time has lapsed. The reported mean concentration of radiocesium in a duplicate diet collected from Fukushima Prefecture in December 2011 was 1.5 Bq kg^{-1} [4], which was lower than that in 2013 as determined in this study. This is because the concentration of radiocesium in foods determined by a duplicate method decreased during processing and cooking and market dilution effect.

The concentrations of $^{134, 137}$Cs and ^{90}Sr were determined in surface soils and agricultural plants collected from both outside and within the difficult-to-return zone in Fukushima Prefecture. The concentration of radiocesium in the agricultural soils outside the zone is decreasing because of plowing, migration, erosion, etc. However, the soil within the difficult-to-return zone (experimental field 5 km west from the

Table 16.3 Concentrations of [134]Cs, ^{137}Cs, [40]K and ^{90}Sr in agricultural plants collected from Fukushima Prefecture in 2012 and 2013

Sample no.	Agricultural plant	Dry matter content (%)	Concentration			
			[134]Cs	[137]Cs	[40]K	[90]Sr
			(Bq kg^{-1} fresh weight)			
AP-1	Komatsuna	4.5	0.030 ± 0.0036	0.055 ± 0.0044	100 ± 0.34	0.054 ± 0.0027
AP-2	Cucumber	4.2	0.063 ± 0.0074	0.11 ± 0.008	66 ± 0.57	0.013 ± 0.0011
AP-3	Brown rice	89	0.74 ± 0.054	1.6 ± 0.077	65 ± 1.9	0.013 ± 0.0018
AP-4	Potato	19	1.7 ± 0.026	3.9 ± 0.039	130 ± 0.88	0.012 ± 0.00093
AP-5	Carrot	9.1	0.36 ± 0.032	0.78 ± 0.040	130 ± 1.7	0.031 ± 0.0022
AP-6	Soy bean	90	3.7 ± 0.32	8.8 ± 0.47	540 ± 14	0.30 ± 0.014
AP-7	Persimmon	14	1.5 ± 0.047	3.6 ± 0.074	56 ± 1.2	0.0086 ± 0.00050
AP-8	Edible chrysanthemum (flowers)	8.8	0.072 ± 0.0040	0.17 ± 0.0059	86 ± 0.32	0.044 ± 0.0039
AP-9	Shiitake mushroom	11	2.2 ± 0.093	5.1 ± 0.14	85 ± 2.3	0.0047 ± 0.00032
AP-10[a]	Pumpkin	15	27 ± 0.79	80 ± 1.3	75 ± 5.9	0.31 ± 0.0061
AP-11[a]	Cabbage	5.8	17 ± 0.38	50 ± 0.68	64 ± 3.5	0.21 ± 0.0057
	Various agricultural plants[b]		ND[c] ~ 4.9	ND ~ 10		ND ~ 0.91

The errors indicate one standard deviation of counting statistics
[a]Collected from the difficult-to-return zone (Okuma)
[b]Samples collected throughout Japan excluding Fukushima Prefecture in 2013 (data from Environmental Radioactivity and Radiation in Japan [6])
[c]Not detected

Nuclear Power Stations in Okuma) is still highly contaminated with radiocesium (Fig. 16.1). The concentration of ^{90}Sr in the soils collected outside the zone is low with no differences among the values. In contrast, ^{90}Sr concentration in the soils collected within the zone was several times higher than that outside the zone. However, the concentration of ^{90}Sr in the soils collected both outside and within the zone is within the range collected throughout Japan except Fukushima (ND-5.9 Bq kg^{-1}, data from Environmental Radioactivity and Radiation in Japan [6]). The concentration of radiocesium in the plants collected outside of the difficult-to-return zone (Table 16.3) is similar to the range shown in Table 16.2. The concentration of radiocesium in the agricultural plants within the zone was still higher than that outside the zone (Table 16.3); however, part of the plants cultivated

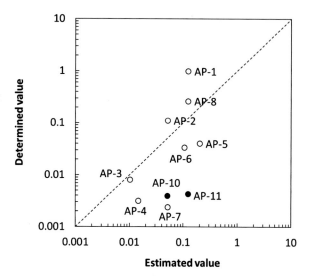

Fig. 16.2 Comparison of ^{90}Sr/^{137}Cs concentration ratio between the estimated value and the determined value in agricultural plants. *Plotted numbers* indicate sample no. in Table 16.2. The *solid-black circles* (AP-10 and AP-11) were collected within the difficult-to-return zone

in the experimental field was lower than the new Standard Limits [7]. The range of ^{90}Sr concentration in the plants collected outside the zone is 0.0047–0.30 Bq kg^{-1} fresh weight, and those collected within the zone were 0.31 (pumpkin) and 0.21 (cabbage). These data are also within the range collected throughout Japan except Fukushima (ND-0.91 Bq kg^{-1} fresh weight, data from Environmental Radioactivity and Radiation in Japan [6]).

The new Standard Limits of radiocesium include the contribution of ^{90}Sr concentration in general foods. The concentration ratio of ^{90}Sr/^{137}Cs in the foods was predicted by using observed ^{90}Sr/^{137}Cs concentration ratio (0.003) [8] in the soils from the FDNPS accident and the reported soil-to-plant transfer factor [9, 10], and the concentration of ^{90}Sr in foods was determined by multiplying the predicted ^{90}Sr/^{137}Cs ratio in foods by the measured ^{137}Cs value in foods. Therefore, the propriety between the predicted and the measured ^{90}Sr/^{137}Cs ratio in foods (Fig. 16.2) needs to be evaluated. The measured concentration ratio of ^{90}Sr/^{137}Cs in the plants, except for three samples (AP-1, komatsuna; AP-2, cucumber; AP-8, edible chrysanthemum), was lower than the predicted ^{90}Sr/^{137}Cs ratio, and the determined ratio in the two samples collected within the difficult-to-return zone (Okuma), which may have a large contribution from the accident, was also lower than the predicted ^{90}Sr/^{137}Cs ratio. The concentration of ^{90}Sr in the soils collected from the three fields, where the ^{90}Sr/^{137}Cs concentration ratio in the plants overestimated the predicted ratio, was similar in range to the global fallout deposited in the soil. Therefore it is necessary to attribute the ^{90}Sr contents in the plants as being derived from the global fallout from several decades ago.

Internal radiation doses from radiocesium through food ingestion for adult males and females (over the age of 19) were estimated. Measured and predicted data for the radiocesium concentration in food categories were used. The concentration of ^{137}Cs in the animal products including milk collected in 2012 and 2013 was not detected, and the average value of the detection limits (0.6 Bq kg^{-1}) in the animal products was used for the dose estimation. Drinking water pathway was not included for the dose estimation because it was lower than the detection limit. The estimated internal radiation doses through food ingestion for males and females were 0.066 and 0.052 mSv y^{-1} in 2012, and those in 2013 were 0.016 and 0.012, respectively, reflecting the decreases in the concentration of radiocesium in foods with time elapsed. It was also reported that the internal radiation dose from radiocesium in Fukushima Prefecture in 2012 was 0.0039–0.0066 mSv y^{-1} by the market basket method [11], which was one order of magnitude lower than that in this study. This is attributed to the fact that the collected foods by the market basket method usually included products both within and outside of Fukushima Prefecture, and the concentration of radiocesium in the foods decreased by market dilution. On the other hand, the samples collected in this study were produced only in Fukushima Prefecture and were not influenced by the market dilution effect. The internal radiation doses from radiocesium by the duplicate diet method in Fukushima Prefecture were reported as 0.026 mSv y^{-1} in 2011 [4] and 0.0022 mSv y^{-1} in 2012 [11]. The internal radiation dose from radiocesium through food ingestion determined by the duplicate diet method is lower than that by the market basket method because of processing and cooking, and it is assumed that those values decrease with time elapsed.

Acknowledgement This work was supported by MHLW KAKENHI Grant. We are grateful to Messrs. A. Sato (Ichii Co. Ltd.) and M. Kanno (Citizens for Revitalization of Oguni after Nuclear Disaster) for sample collection, and Dr. P. T. Lattimore for his useful suggestion and comments. We thank Mr. A. Kanno, Mses. C. Suzuki, W. Horiuchi and M. Kato for sample pretreatment.

References

1. Tsukada H, Takeda A, Takahashi T, Hasegawa H, Hisamatsu S, Inaba J (2005) Uptake and distribution of ^{90}Sr and stable Sr in rice plants. J Environ Radioact 81:221–231
2. Takeda A, Tsukada H, Nakao A, Takaku Y, Hisamatsu S (2013) Time-dependent changes of phytoavailability of Cs added to allophanic Andosols in laboratory cultivations and extraction tests. J Environ Radioact 122:29–36
3. Tsukada H (2014) Behavior of radioactive cesium in soil with aging. Jpn J Soil Sci Plant Nutr 85:77–79 (in Japanese)
4. Harada KH, Fujii Y, Adachi A, Tsukidate A, Asai F, Koizumi A (2013) Dietary intake of radiocesium in adult residents in Fukushima Prefecture and neighboring regions after the Fukushima nuclear power plant accident: 24-h food-duplicate survey in December 2011. Environ Sci Tech 47:2520–2526
5. Nuclear Regulation Authority. Monitoring information of environmental radioactivity level. http://ramap.jmc.or.jp/map/eng/

6. Environmental Radioactivity and Radiation in Japan. http://www.kankyo-hoshano.go.jp/en/index.html
7. Ohse K, Kitayama K, Suenaga S, Matsumoto K, Kanno A, Suzuki C, Kawatsu K, Tsukada H (2014) Concentration of radiocesium in rice, vegetables, and fruits cultivated in evacuation area in Okuma Town, Fukushima. J Radioanal Nucl Chem 303:1533–1537
8. Ministry of Health, Labour and Welfare. http://www.mhlw.go.jp/stf/shingi/2r98520000023nbs-att/2r98520000023ng2.pdf
9. IAEA (2010) Handbook of parameter values for the prediction of radionuclide transfer in terrestrial and freshwater environment, vol 472, Technical report series. International Atomic Energy Agency, Vienna
10. Tsukada H, Nakamura Y (1998) Transfer factors of 31 elements in several agricultural plants collected from 150 farm fields in Aomori, Japan. J Radioanal Nucl Chem 236:123–131
11. Ministry of Health, Labour and Welfare. http://www.mhlw.go.jp/shinsai_jouhou/shokuhin.html

Chapter 17
Analysis of Factors Causing High Radiocesium Concentrations in Brown Rice Grown in Minamisoma City

Takashi Saito, Kazuhira Takahashi, Toshifumi Murakami, and Takuro Shinano

Abstract Despite a concentration of exchangeable K of >208 mg kg^{-1} dry weight in soil, the brown rice grown in Minamisoma City in 2013 had a higher concentration of radiocesium than the new Japanese standard (100 Bq kg^{-1}) for food. To analyze the factors affecting the radiocesium concentration in brown rice, we carried out pot tests using paddy soil and irrigation water collected in Minamisoma City. Rice seedlings were planted in 5-L pots containing Minamisoma soil, in which the exchangeable K was 125 mg kg^{-1} dry weight, and were irrigated with tap water or irrigation water collected in Minamisoma City. There was no difference in the Cs-137 concentration in brown rice between the two types of irrigation. Then we grew rice in the Minamisoma soil and two soils collected in Nakadori, Fukushima Prefecture. Cs-137 uptake in the Minamisoma soil was intermediate between the uptake rates in the Nakadori soils, showing that the Minamisoma soil was not special in radiocesium uptake. Finally, we grew rice in soil without radiocesium near the Fukushima Daiichi Nuclear Power Plant in 2014. Although the maximum value of Cs-137 in brown rice was 18 Bq kg^{-1}, below the standard, radiocesium was attached to the surface of the foliage.

Keywords Cs-137 • Brown rice • Exchangeable K • Irrigation water • Minamisoma City

T. Saito (✉) • K. Takahashi
Agro-environment Division, Fukushima Agricultural Technology Centre, 116 Shimonakamichi, Takakura-aza, Hiwada-machi, Koriyama, Fukushima 963-0531, Japan
e-mail: saito_takashi_01@pref.fukushima.lg.jp

T. Murakami • T. Shinano
NARO Tohoku Agricultural Research Centre, 50 Harajukuminami, Arai Fukushima, Fukushima 960-2156, Japan

© The Author(s) 2016
T. Takahashi (ed.), *Radiological Issues for Fukushima's Revitalized Future*, DOI 10.1007/978-4-431-55848-4_17

17.1 Introduction

Following the accident at the Fukushima Daiichi Nuclear Power Plant on 11 March 2011, soils became contaminated with radiocesium (Cs-134 and Cs-137). To reduce the uptake of radiocesium by rice, growers have been adding potassium (K) fertilizer to their paddy fields, as the uptake of Cs decreases with increasing K concentration ([K]) in the soil [1] and the concentration of radiocesium in brown rice decreases at increasing concentrations of exchangeable K in the soil and K^+ in the soil solution [2]. K fertilization offers an effective and practical way to reduce radiocesium uptake by rice from several soil types [3]. Following the accident, the Food Sanitation Law of 2012 reduced the standard for the concentration of radiocesium in food to 100 Bq kg^{-1}. A concentration of >208 mg kg^{-1} dry weight of exchangeable K in soil is recommended for keeping the radiocesium content in brown rice below the standard [4].

However, despite a concentration of exchangeable K of >208 mg kg^{-1} in soil, the brown rice grown in Minamisoma City in 2013 exceeded the new standard [5].

One possible source is irrigation water. Dissolved radiocesium moves more easily into plants from water than from soil [6]. The concentration of dissolved radiocesium in irrigation water drawn from the Ota River in Minamisoma City was higher than that in other parts of Fukushima Prefecture [7].

One possible source is soil. Although Tsumura et al. [8] reported the relationship between the concentration of exchangeable K in soils and Cs-137 uptake in brown rice, however, discussion on radiocesium uptake in brown rice has not been done in same levels of exchangeable K in soil.

Another possible source is dust. The Cs-137 concentration ([Cs-137]) in dust collected in Futaba Town on 19 August 2013 was clearly higher than that at other times [9].

To identify the cause of the high radiocesium concentration in brown rice grown in Minamisoma City, we conducted pot experiments comparing sources of irrigation water and soil types. In addition, we determined the [Cs-137] of rice grown near the nuclear plant.

17.2 Materials and Methods

17.2.1 Irrigation Water

We collected 20 L of water on six dates (shown in Fig. 17.1) from the Ota River and passed it through 0.45-μm filters. Suspended matter collected on the filters was compressed into cylindrical polystyrene containers (i.d. 5.0 cm, o.d. 5.6 cm, height 6.8 cm) for analysis. The filtrates were concentrated to 2 L by heat and then placed in 2-L Marinelli beakers for analysis as described below. The concentrations of dissolved Cs-137 were 0.15–0.42 Bq L^{-1}, and those of suspended Cs-137 were 0.09–0.17 Bq L^{-1} (Fig. 17.1). [Cs-137] in tap water was 0.02 Bq L^{-1}.

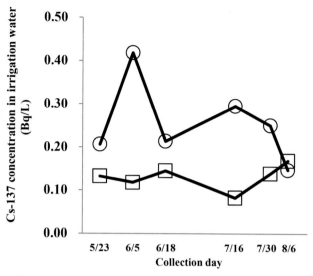

O Dissolved Cs-137 concentration ; □ Suspended Cs-137 concentration

Fig. 17.1 [Cs-137] in irrigation water used for pot experiments. Error bars represent standard errors ($n = 3$). *$P < 0.05$ (Student's t-test) between tap water and irrigation water

Table 17.1 Chemical properties of soils used for pot experiment

Name	Soil types	pH(H₂O)	Total C (g kg⁻¹)	K (mg kg⁻¹)	Exchangeable Ca	Mg	CEC (cmolc kg⁻¹)
Minamisoma	Gray lowland soil	6.3	14	150	1640	390	9.1
Soil A	Andosol	5.6	79	25	1220	170	15.2
Soil B	Gray lowland soil	6.6	9.1	134	2160	553	11.4

17.2.2   Soils

Soils were collected from the field in April 2014. Experiment 1 used a contaminated Glay lowland soil collected from the top 15 cm of a paddy field in Minamisoma City where the concentration of radiocesium in brown rice grown in 2013 exceeded 100 Bq kg⁻¹.

Experiment 2 used three soils (Table 17.1): the Minamisoma Gray lowland soil; an Andosol collected from 0 to 15 and 15–30 cm depth in a paddy field in northern Fukushima Prefecture where the concentration of radiocesium in brown rice grown in 2011 exceeded 500 Bq kg⁻¹ (soil A); and a Gray lowland soil collected from 0 to 15 and 15–30 cm depth in a paddy field at the Fukushima Agricultural Technology Centre (FATC), where the concentration of radiocesium in brown rice grown in 2011 was below the limit of quantification (<20 Bq kg⁻¹) (soil B).

Experiment 3 used an uncontaminated Gray lowland soil collected from the subsoil (beneath 15 cm depth) of a paddy field at FATC. The [Cs-137] in the soil was 35 Bq kg^{-1} dry weight.

17.2.3 Pot Experiments

Soils were air dried, thoroughly mixed, and passed through a 2-mm sieve. On 8 May 2014, rice seeds (*Oryza sativa* L. 'Maihime') were sown in granular culture soil. On 6 June, four seedlings were transplanted into each 5-L Wagner pot (diam. 16 cm, height 25 cm), which held 3.0 kg dry weight of soil. Each pot also received a basal dressing of 1.0 g of ammonium sulfate and calcium superphosphate and mixed into the soil. The water level was kept at a depth of 3–5 cm during the experiments. All experiments were conducted at FATC (and experiment 3 at Okuma Town also) under natural light. All treatments had three replicates. The rice plants were harvested on 9 October 2014, and samples of brown rice and leaves were oven dried at 40 °C for 24 h.

17.2.3.1 Experiment 1

The [Cs-137] in the contaminated Minamisoma soil was 1500 Bq kg^{-1} dry weight. No KCl was applied. Pots were watered with either tap water or irrigation water. Each pot received a total of 11–16 L during the experiment.

17.2.3.2 Experiment 2

The [Cs-137] in the contaminated Minamisoma soil was adjusted to 1500 Bq kg^{-1} as above. The [Cs-137] in soils A and B was adjusted to 1500 Bq kg^{-1} by mixing the topsoil and subsoil. The exchangeable K content of each was adjusted to 125 or 250 mg kg^{-1} with KCl. Pots were watered with irrigation water used in Experiment 1.

17.2.3.3 Experiment 3

The exchangeable K content of the uncontaminated soil was adjusted to 208 mg kg^{-1} with KCl. All plants were watered with tap water. Treatment pots were moved from the FATC to Okuma Town on 9 July and back to the FATC on 20 August. Control pots remained at the FATC. The pots were protected with a multi-film so that only the leaves were exposed to fallout. Radiocesium contaminations of rice foliage were analyzed by a gamma-ray spectrometry and autoradiography visually. In the autoradiogram analysis, powdered shoot (3 g) was put into a small polyethylene bag (10 × 7 cm) and the bag was put on the cardboard (40 × 20 cm, 0.6 mm thickness).

Markers made with potassium chloride (contain 10–26 mg) were attached on the corners of the cardboard samples to obtain a superposition of the autoradiogram and the visible image. An imaging plate (BAS-SR2040 (40×20 cm), Fuji-film, Japan) was contacted with the cardboard, and they were put into a paper case together and sandwiched between two lead plates of 4 mm thickness in a dark room. After 7 days exposure the imaging plate was scanned by image scanner (Typhoon FLA 7000, GE Healthcare Bio-Science Co., Ltd., USA) at a spatial resolution of 25 μm. The autoradiography and its visible image were overlapped on image processing software (Photoshop CS4 ver. 11.0, Adobe Co., Ltd., USA).

17.2.4 Soil and Plant Analyses

The chemical properties of the soils were analyzed according to the Editorial Boards of Methods for Soil Environment Analysis [10] (Table 17.1). Soil pH (H_2O) was measured at a soil-to-water ratio of 1:2.5 (w/w). The total carbon content was determined by dry combustion on a Sumigraph NC Analyzer NC-220 F (Sumika Chemical Analysis Service, Ltd., Osaka, Japan). Exchangeable K, calcium, magnesium were determined by the semi-micro Schollenberger method on an atomic absorption spectrophotometer (AA280FS; Varian Technologies Japan Ltd., Tokyo, Japan), and cation exchange capacity is calculated as the sum of these component ions.

The brown rice and leaf samples were compressed into cylindrical polystyrene containers as above, and the [Cs-137] was measured with a Ge gamma-ray detector connected to a multichannel analyzer (GC2020, GC3020, GC3520, GC4020, Canberra USA) for 36,000 s.

17.2.5 Statistical Analyses

Statistical analyses were performed in StatView 5.0 J software (SAS Institute, Berkeley, CA, USA). Analysis of variance (ANOVA) followed by t-test or Tukey's multiple comparison test was used to determine the significance of differences in a pairwise comparison matrix.

17.3 Results

17.3.1 Effect of Irrigation Water on Cs-137 Uptake in Rice

In both watering treatments (irrigation and tap water), the concentration of exchangeable K was 125 mg kg^{-1} before planting and 45 mg kg^{-1} after harvest (data not shown). When watered with irrigation water, [Cs-137] in brown rice and

Fig. 17.2 [Cs-137] in foliage and brown rice of plants watered with irrigation or tap water. Error bars represent standard error ($n = 3$). ns, not significantly different; $*P > 0.05$ (Student's t-test) between irrigation water and tap water

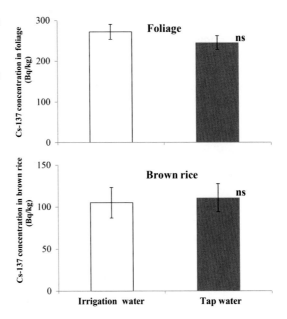

Both treatments : Exchangeable K was 125 mg kg^{-1} before planting, and 45 mg kg^{-1} after harvest

foliage were 105 and 272 Bq kg^{-1} and watered with tap water were 111 and 245 Bq kg^{-1}, respectively (Fig. 17.2). There was no significant difference in the [Cs-137] of brown rice or foliage between treatments.

17.3.2 Effect of Soil Type on Cs-137 Uptake in Rice

When exchangeable K in Minamisoma soil of pot was 34.9 and 52.1 mg kg^{-1} after harvest, the [Cs-137] in the foliage of rice plant grown were 272 and 119 Bq kg^{-1}, respectively (Fig. 17.3). On the other hand, when exchangeable K in soil A of pot was 15.6 and 46.5 mg kg^{-1} after harvest, the [Cs-137] in the brown rice grown was 594 and 215 Bq kg^{-1}, respectively, and higher than that of Minamisoma soil. When exchangeable K in soil B of pot was 56.4 and 78.8 mg kg^{-1}, the [Cs-137] in the brown rice was 30.4 and 12.0 Bq kg^{-1}, respectively, and lower than that of the Minamisoma soil.

When exchangeable K in Minamisoma soil of pot was 34.9 and 52.1 mg kg^{-1} after cultivating, the [Cs-137] in the brown rice grown were 105 and 54.1 Bq kg^{-1}, respectively. On the other hand, when exchangeable K in soil A of pot was 15.6 and 46.5 mg kg^{-1}, the [Cs-137] in the brown rice grown was 206 and 93.8 Bq kg^{-1},

Fig. 17.3 Relationship between [Cs-137] in rice and exchangeable K after harvest in different kind soils or in different K soils. Error bars represent standard errors ($n = 3$)

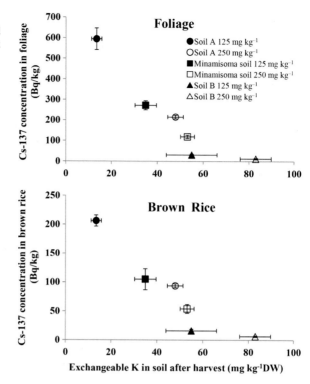

respectively, and higher than that of the Minamisoma soil. When exchangeable K in soil B of pot was 56.4 and 78.8 mg kg^{-1}, the [Cs-137] in the brown rice was 15.8 and 6.3 Bq kg^{-1}, respectively, and lower than those of Minamisoma soil.

17.3.3 Effect of Site on Acquisition of Cs-137 by Foliage and Brown Rice

There was no significant change in the amount of exchangeable K in either treatment before and after the experiment (data not shown). The [Cs-137] in the brown rice and leaves of plants grown at Okuma Town for 6 weeks was 5–7 times that in the plants kept at the FATC (Fig. 17.4).

Autoradiographs of brown rice and foliage revealed no contamination of either in plants grown at the FATC or in brown rice of plants grown at Okuma Town (Fig. 17.5). However, foliage of plants grown at Okuma Town showed radioactive contamination (Fig. 17.6).

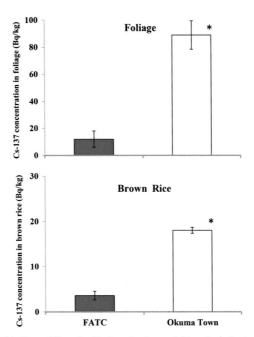

FATC : Exchangeable K was 208 mg kg⁻¹ before planting, and 63 mg kg⁻¹ after harvest.

Okuma Town: Exchangeable K was 208 mg kg⁻¹ before planting, and 67 mg kg⁻¹ after harvest.

Fig. 17.4 Translocation of Cs-137 from foliage to brown rice between two sites. Error bars represent standard errors ($n = 3$). *$P < 0.05$ (Student's t-test) between TATC and Okuma Town

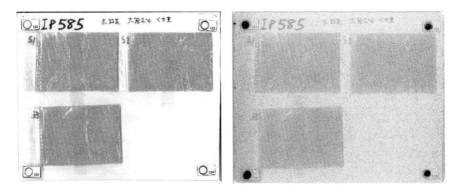

Fig. 17.5 Autoradiographs of foliage rice by FATC

Fig. 17.6 Autoradiographs of foliage rice by Okuma Town

17.4 Discussion

17.4.1 Effect of Irrigation Water on Cs-137 Uptake in Rice

In hydroponic culture with 0.1, 1.0, or 10 Bq L^{-1} Cs-137, the [Cs-137] in brown rice increased with increasing [Cs-137] in the culture solution [6]. However, nutrient uptake can be greater in hydroponic culture than in soil culture because all nutrients are in solution. In contrast, a high level of exchangeable K in soil can limit the transfer of Cs-137 from irrigation water containing low levels of Cs-137 (0.1–1.0 Bq L^{-1}) [11]. We found no significant difference in uptake between water sources; therefore, low levels of exchangeable K in both soils limited Cs-137 uptake by rice plants.

17.4.2 Effect of Soil Type on Cs-137 Uptake in Rice

The Minamisoma soil did not have superior ability to promote Cs-137 uptake (Fig. 17.3). To reduce the [Cs-137] below 100 Bq kg^{-1} would require 40 mg kg^{-1} of exchangeable K in the Minamisoma soil and 50 mg kg^{-1} in soil A. Thus, at a similar level of exchangeable K in the soil, Cs-137 uptake depended on soil type. The high carbon content and low clay content of soil A may have helped to inhibit Cs-137 uptake. Therefore, the Minamisoma soil was not the cause of high [Cs-137] in 2013.

17.4.3 Effect of Site on Acquisition of Cs-137 by Foliage and Brown Rice

The [Cs-137] of rice plants increased greatly during 6 weeks' culture in Okuma Town. Cs-137 derived from the Fukushima accident is still distributed widely around the town. Thus, the rice plants could have taken up more radiocesium from the outside environment.

References

1. Tsukada H, Hasegawa H, Hisamatsu S, Yamasaki S (2002) Transfer of ^{137}Cs and stable Cs from paddy soil to polished rice in Aomori, Tsukuba, Japan. J Environ Radioact 59:351–363
2. Saito T et al (2012) Effect of potassium application on root uptake of radiocesium in rice. In: Proceeding of international system on environmental monitoring and dose estimation of residents after accident of TEPCO's Fukushima Daiichi Nuclear Power Station. Kyoto University Research Reactor Institute Press, Kyoto, pp 165–169
3. Kato (2012) Countermeasures to reduce radiocaesium contamination I paddy rice, soy bean and cabbage. In: International science symposium on combating radionuclide contamination in Agro-soil environment, Fukushima, pp 317–318
4. NARO (2012) http://www.naro.affrc.go.jp/org/tarc/seika/jyouhou/H24/kankyou/H24kankyou012.html (in Japanese, May 2015)
5. MAFF (2014a) http://www.maff.go.jp/j/kanbo/joho/saigai/fukusima/pdf/25kome_h26_01.pdf (in Japanese, May 2015)
6. Nemoto K, Abe J (2013) Radiocesium absorption by rice in paddy field ecosystem. In: Nakanishi TM, Tanoi K (eds) Agricultural implications of the Fukushima nuclear accident. Springer, Tokyo, pp 19–27
7. MAFF (2014b) http://www.maff.go.jp/j/kanbo/joho/saigai/fukusima/pdf/yousui_h26_8.pdf (in Japanese, May 2015)
8. Tsumura A et al (1984) Behavior of radioactive Sr and Cs in soils and soil-plant systems. Nat Inst Agro-Environ Sci Rep 36:57–113
9. Tsuruta et al (2014) Long-term changes for the three years of the radioactive material concentration of atmospheric aerosols in Fukushima and its surrounding 3 point. In: Abstracts of the 15th workshop on environmental radioactivity, p1
10. Editorial Boards of Methods for Soil Environment Analysis (1997) Methods for soil environment analysis. Hakubunkan Shinsha Publishers Press, Tokyo
11. Suzuki Y et al (2015) Effect of the concentration of radiocesium dissolved in irrigation water on the concentration of radiocesium in brown rice. Soil Sci Plant Nutr 61:191–199

Chapter 18
Radiocesium and Potassium Decreases in Wild Edible Plants by Food Processing

Keiko Tagami and Shigeo Uchida

Abstract It is more than 4 years since March 11, 2011, and, at this stage, foods that exceed the standard limits of radiocesium are mainly from the wild. Hence, one of the public's main concerns is how to decrease ingestion of radiocesium from foods they have collected from the wild as well as from their home-grown fruits because radioactivities in these food materials have not been monitored. In this study, we focused on wild edible plants and fruits, and the effects of washing, boiling, and pealing to remove radiocesium were observed. Samples were collected in 2013 and 2014 from Chiba and Fukushima Prefectures, e.g., young bamboo shoots, giant butterbur, and chestnuts. Wild edible plants were separated into three portions to make raw, washed, and boiled samples. For fruit samples (i.e., persimmon, loquat, and Japanese apricot), fruit parts were separated into skin, flesh, and seeds.

It was found that washing of plants is not effective in removing both ^{137}Cs and ^{40}K, and that boiling provided different removal effects on plant tissues. The retention factors of ^{137}Cs and ^{40}K for thinner plant body sample (leaves) tended to be higher than those for thicker plant body types, e.g., giant butterbur petiole and bamboo shoots. Thus, the boiling time as well as the crop thickness affects radiocesium retention in processed foods. For fruits, Cs concentration was higher in skin than in fruit flesh for persimmon and loquat; however, Japanese apricot showed different distribution.

Keywords Food processing retention factor • Radiocesium • Potassium • Wild edible plants • Fruits

18.1 Introduction

Radiocesium (^{134}Cs $+^{137}$Cs) concentrations in foods are of great concern in Japan since the Fukushima Daiichi Nuclear Power Plant (FDNPP) accident, as people wish to avoid receiving additional internal doses from ingestion of the radionuclide.

K. Tagami (✉) • S. Uchida
National Institute of Radiological Sciences, Anagawa 4-9-1, Inage-ku, Chiba-shi, Chiba 263-8555, Japan
e-mail: k_tagami@nirs.go.jp

© The Author(s) 2016
T. Takahashi (ed.), *Radiological Issues for Fukushima's Revitalized Future*,
DOI 10.1007/978-4-431-55848-4_18

Food monitoring has been carried out since March 2011; provisional regulation values for total radiocesium concentration were applied at the time, and the values were 200 Bq kg^{-1} for water, milk, and daily products, and 500 Bq kg^{-1} for other food materials. From April 1, 2012, standard values for radiocesium have been employed in Japan, and the concentration in marketed raw food materials has been limited up to 100 Bq kg^{-1}, except baby foods (50 Bq kg^{-1}) and drinking water (10 Bq kg^{-1}). If any food exceeds these limits, the food name together with the producing district has been reported immediately by the Ministry of Health, Labour, and Welfare (MHLW). Every month, radioactivities of more than 20,000 samples are being measured and the recent data of February 2015 [1] have shown that foods exceeding the standard limits were mostly from the wild (i.e., not commercially grown or raised), i.e., meats of wild boar, Sika deer, Asian black bear, Japanese rock fish (marine), Japanese eel and char (freshwater). Not only meats but also edible wild plants in spring and autumn in 2014, e.g., giant butterbur flower-bud, Japanese angelica-tree shoot (*taranome*), and various mushrooms species, exceeded the limits.

One of the public's main concerns is how to decrease ingestion of radiocesium from foods, especially foods they have collected from the wild as well as from their home-grown garden fruits because radioactivities in these foods have not been measured for many cases. Unfortunately, radiocesium removal data by food processing, including culinary preparation, for crops commonly consumed in Japan have been limited because of little interest in the topic before the FDNPP accident. To remedy this, Japanese researchers have begun collecting such data [2–12]. Recently, the Radioactive Waste Management Funding and Research Center (RWMC) compiled the radiocesium removal ratios by food processing using open source data published mainly in 2011 and 2012 [13]. However, it is necessary to add more data to update the information.

In the present study, we focused on food processing effects on radiocesium removal in food plants from the wild to add more information and we compared obtained values with previously compiled values in the IAEA Technical Report Series No. 472 [14]. We also measured potassium-40 (^{40}K) concentrations in the same samples for comparison with radiocesium.

18.2   Materials and Methods

The following samples were collected in Chiba and Fukushima Prefectures in 2013 and 2014: giant butterbur (*Petasites japonicus*: flower-bud, 2; leaf blade, 4; petioles, 4), Japanese mugwort (*Artemisia indica var. Maximowiczii*: young shoots, 4), field-horsetail (*Equisetum arvense*: fertile stem, 3), water dropwort (*Oenanthe javanica*: young shoots, 1), Moso bamboo (*Phyllostachys heterocycla f. pubescens*: young shoots, 8), chestnut (*Castanea crenata*: nuts, 3), persimmon (*Diospyros kaki*: fruits, 2), loquat (*Eriobotrya japonica*: fruits, 4), and Japanese apricot (*Prunus mume*, 1). Sampling dates are listed in Tables 18.1, 18.2, and 18.3. Immediately after the

Table 18.1 Food processing retention factor (F_r) of ^{137}Cs and ^{40}K for giant butterbur tissues collected in 2013–2014

Tissue	Sampling date	Method of processing	Wet mass ratio	F_r of ^{137}Cs	F_r of ^{40}K
Flower bud	2013/3/13	Boiling	1.1	0.19 ± 0.11	0.56 ± 0.10
Flower bud	2013/3/18	Boiling	1.2	0.40 ± 0.18	0.65 ± 0.14
Leaf blade	2013/4/5	Washing	1.0	0.90 ± 0.10	1.24 ± 0.08
Leaf blade	2013/4/19	Washing	1.0	1.01 ± 0.08	1.10 ± 0.07
Leaf blade	2013/4/30	Washing	1.0	1.04 ± 0.08	0.99 ± 0.05
Leaf blade	2014/4/8	Washing	1.0	0.98 ± 0.10	0.99 ± 0.04
Leaf blade[a]	*2012–2014*	*Washing*	–	*0.95 (n = 8)*	*1.02 (n = 8)*
Leaf blade	2013/4/5	Boiling	1.1	0.43 ± 0.05	0.51 ± 0.04
Leaf blade	2013/4/19	Boiling	1.0	0.39 ± 0.03	0.60 ± 0.04
Leaf blade	2013/4/30	Boiling	1.0	0.46 ± 0.04	0.56 ± 0.03
Leaf blade	2014/4/8	Boiling	0.95	0.40 ± 0.05	0.44 ± 0.02
Leaf blade[a]	*2012–2014*	*Boiling*	–	*0.40 (n = 8)*	*0.46 (n = 8)*
Petiole	2013/4/5	Washing	1.0	0.69 ± 0.16	1.20 ± 0.07
Petiole	2013/4/19	Washing	1.0	1.34 ± 0.26	0.90 ± 0.05
Petiole	2013/4/30	Washing	1.0	1.10 ± 0.18	1.01 ± 0.05
Petiole	2014/4/8	Washing	1.0	0.91 ± 0.20	0.94 ± 0.04
Petiole[a]	*2012–2014*	*Washing*	–	*0.97 (n = 8)*	*1.01 (n = 8)*
Petiole	2013/4/5	Boiling	0.83	0.59 ± 0.12	0.82 ± 0.05
Petiole	2013/4/19	Boiling	0.84	0.93 ± 0.18	0.73 ± 0.04
Petiole	2013/4/30	Boiling	0.84	0.66 ± 0.10	0.66 ± 0.03
Petiole[b]	2014/4/8	Boiling	1.0	0.37 ± 0.10	0.35 ± 0.02
Petiole[a]	*2012–2014*	*Boiling*	–	*0.80 (n = 8)*	*0.80 (n = 8)*

± shows error from counting
[a] Averaged value for the data collected from 2012 (published in [2]) to 2014
[b] The sample was kept in water for 1 h after boiling

collection, samples were transferred to a laboratory and weighed to obtain the fresh weight.

In order to obtain the food processing effect, giant butterbur, Japanese mugwort, and water dropwort samples were separated into three portions to make raw, washed, and boiled (2.5 min) subsamples. One giant butterbur petiole sample was soaked in water for 1 h at room temperature after boiling. Field horsetail, young bamboo shoot, and chestnut samples were separated into two portions to make raw and boiled subsamples (boiling times depended on samples). All samples were weighed before processing. Washing was carried out with tap water in a washing bowl by changing the water five times, and then, finally, the samples were rinsed with reverse osmosis (RO) water. For the boiling process, edible parts of the plants were cooked in RO water after washing with tap water five times. For fruits, a whole fruit was washed with running tap water and rinsed with RO water. Then the water was removed with paper towels from the fruits, and each tissue part (skin, flesh, and seeds) was separated and weighed.

Table 18.2 Food processing retention factor (F_r) of ^{137}Cs and ^{40}K for eight bamboo shoots collected in 2013

Tissue	Sampling date	Method of processing	Wet mass ratio	F_r of ^{137}Cs	F_r of ^{40}K
New shoot-1	2013/4/9	Boiling	0.97	0.94 ± 0.04	0.80 ± 0.05
New shoot-2	2013/4/9	Boiling	0.95	0.66 ± 0.04	0.67 ± 0.06
New shoot-3	2013/4/9	Boiling	0.95	0.65 ± 0.02	0.55 ± 0.06
New shoot-4	2013/4/9	Boiling	–	0.68 ± 0.03	0.69 ± 0.05
New shoot-5	2013/4/9	Boiling	–	0.65 ± 0.03	0.69 ± 0.04
New shoot-6	2013/4/9	Boiling	0.99	0.67 ± 0.02	0.63 ± 0.06
New shoot-7	2013/4/9	Boiling	0.97	0.88 ± 0.04	0.75 ± 0.06
New shoot-8	2013/4/9	Boiling	0.97	0.79 ± 0.05	0.80 ± 0.07
New shoot[a]	*2012–2013*	*Boiling*		*0.71 (n = 12)*	*0.73 (n = 12)*

± shows error from counting
[a] Averaged value for the data collected from 2012 (published in [2]) to 2013

Table 18.3 Food processing retention factor (F_r) of ^{137}Cs and ^{40}K for mugwort (M), field horsetail (F), water dropwort (W), and chestnut (C) collected in 2013–2014

Tissue	Sampling date	Method of processing	Wet mass ratio	F_r of ^{137}Cs	F_r of ^{40}K
M, New shoots	2013/5/8	Washing	1.0	0.87 ± 0.37	1.14 ± 0.07
M, New shoots	2013/5/24	Washing	1.0	1.09 ± 0.02	0.86 ± 0.13
M, New shoots	2013/9/26	Washing	1.0	0.89 ± 0.03	0.97 ± 0.32
M, New shoots	2014/12/9	Washing	1.0	0.77 ± 0.36	0.97 ± 0.05
M, New shoots[a]	*2012–2014*	*Washing*	–	*0.98 (n = 9)*	*0.97 (n = 9)*
M, New shoots	2013/5/8	Boiling	1.0	0.36 ± 0.21	0.57 ± 0.04
M, New shoots	2013/5/24	Boiling	1.1	0.25 ± 0.01	0.45 ± 0.08
M, New shoots	2013/9/26	Boiling	1.0	0.47 ± 0.02	0.64 ± 0.19
M, New shoots	2014/12/9	Boiling	0.94	0.29 ± 0.20	0.26 ± 0.02
M, New shoots[a]	*2012–2014*	*Washing*	–	*0.39 (n = 8)*	*0.46 (n = 8)*
F, Fertile stem	2013/3/8	Boiling	0.81	0.62 ± 0.29	0.43 ± 0.05
F, Fertile stem	2013/3/20	Boiling	0.86	0.47 ± 0.25	0.57 ± 0.08
F, Fertile stem	2013/3/26	Boiling	0.75	0.16 ± 0.33	0.44 ± 0.04
F, Fertile stem[a]	*2012–2014*	*Boiling*	–	*0.46 (n = 6)*	*0.52 (n = 6)*
W, New shoots	2013/5/24	Washing	1.0	0.90 ± 0.06	0.96 ± 0.22
W, New shoots	2013/5/24	Boiling	0.83	0.61 ± 0.04	0.32 ± 0.12
C, nuts-1[b]	2013/9/26	Boiling	1.0	1.30 ± 0.01	1.04 ± 0.17
C, nuts-2[b]	2013/9/26	Boiling	1.0	0.68 ± 0.01	0.78 ± 0.15
C, nuts-3[b]	2013/9/26	Boiling	1.0	0.81 ± 0.01	0.77 ± 0.19

± shows error from counting
[a] Averaged value for the data collected from 2012 (published in [2]) to 2014
[b] Samples taken from different trees

Then, all samples were oven-dried at 80 °C to decrease the sample volume. Each oven-dried sample was pulverized and mixed well, and then transferred to a plastic container (U8 container). Radioactivity concentration in each sample was measured by a Ge detecting system (Seiko EG&G) and the gamma spectrum was analyzed using Gamma Station software (Seiko EG&G) to obtain activity on wet mass basis (Bq kg^{-1}-wet mass). The detection limit was about 0.5–1.0 Bq kg^{-1}-wet mass with the counting time of 40,000–80,000 s. The radiocesium concentrations in samples were usually low and sometimes ^{134}Cs could not be detected 2–3 years after the accident. Therefore, in this study, only ^{137}Cs data are presented together with ^{40}K.

Food processing retention factor (F_r) was determined by using the following equation as defined in IAEA TRS-472 [14]:

$$F_r = A_{after} \, (Bq) \, / \, A_{before} \, (Bq)$$

where A_{after} is the total activity of ^{137}Cs or ^{40}K retained in the food after processing (Bq), and A_{before} is the total activity in the food before processing (Bq). The wet mass of the processed subsample before processing (W_{bp}, kg) was recorded, thus, A_{before} was calculated using the following equation:

$$A_{before} = C_{raw} \times W_{bp}$$

where C_{raw} is the radioactivity concentration of ^{137}Cs or ^{40}K in the raw subsample (Bq kg^{-1}-wet mass).

For fruit samples, distribution percentages of wet mass, ^{137}Cs and ^{40}K in skin, flesh, and seeds were calculated and these distributions were compared.

18.3   Results and Discussion

The F_r values of ^{40}K and ^{137}Cs for each wild edible plant sample are shown in Tables 18.1, 18.2, and 18.3. Some F_r data exceeded 1.0 although that is impossible from the above equation; however, because of the low concentrations in samples giving a large counting error, and the samples for raw and processed samples being not completely the same, values of more than 1.0 were inevitable.

The mean F_rs of ^{137}Cs by washing for butterbur leaf blade ($n = 4$), petiole ($n = 4$), mugwort ($n = 4$), and water dropwort ($n = 1$) were 0.98, 1.0, 0.91, and 0.90, respectively and that of ^{40}K were 1.1, 1.0, 0.98, and 0.96, respectively. From this result, it was clear that washing plants is not effective to remove both ^{40}K and ^{137}Cs, because both elements are distributed inside the plant body. Boiling provided different removal effects by plant tissues. The mean F_rs of ^{137}Cs for leaf-blade of giant butterbur ($n = 4$), mugwort ($n = 4$), water dropwort ($n = 1$), and fertile stem of field horsetail ($n = 3$) were 0.40, 0.41, 0.61, and 0.42, respectively, i.e., about 40–60 % of the ^{137}Cs was removed by this process. Similar results were observed for ^{40}K for these samples. Unfortunately, however, the number of samples treated in 2013–2014 was not enough for statistical analysis between F_rs of ^{40}K and ^{137}Cs in each plant species. Only analytical results for bamboo shoots ($n = 8$) are reported here; when ANOVA test was carried out, no statistical difference was observed between F_rs of ^{40}K and ^{137}Cs. For some sample types, all the data from 2012 (reported in [2]) to 2014 were summarized (Tables 18.1, 18.2, and 18.3) and the values for ^{40}K and ^{137}Cs did not show any statistical differences by ANOVA test. Thus, we concluded that K could be used as an analogue for radiocesium to calculate F_rs.

Compared to leafy samples, bamboo shoots and petioles of giant butterbur (except soaking in water for 1 h) showed slightly higher F_rs of ^{137}Cs of 0.74 and 0.73, respectively; since these tissues are thicker than leaf tissues, it is reasonable to expect that more radiocesium would be retained in these tissues. However, when the giant butterbur sample soaked in water for 1 h showed lower F_r values, therefore, boiling followed by soaking would effectively remove ^{137}Cs, although K was also removed by this process. For chestnut, mean values of F_r of ^{137}Cs and ^{40}K were 0.95 and 0.87, respectively. Removal of these two elements by boiling of chestnut was difficult, because the hard shell prevented extracting ^{137}Cs and ^{40}K from the nuts into the boiling water. In total, most of the F_r values presented in this study and our previous data [2] were within the range of values compiled by IAEA [14], as shown in Fig. 18.1.

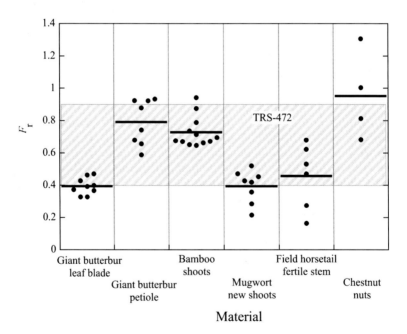

Fig. 18.1 Comparison of food processing retention factor (F_r) of ^{137}Cs for several plant materials collected in 2011–2014 by boiling. Both previous [2] and present study's data are plotted. The *shaded area* shows Fr range for vegetables, berries, and fruits boiled in water reported in TRS-472 by IAEA [14]

Table 18.4 Relative proportions (%) of wet mass, ^{137}Cs and ^{40}K in fruit tissues (skin, flesh, and seeds) to whole fruits for four species collected in 2013–2014 at harvest

Sample	Sampling date	Wet mass, %			^{137}Cs, %			^{40}K, %		
		Skin	Flesh	Seeds	Skin	Flesh	Seeds	Skin	Flesh	Seeds
Persimmon[a]	2011/10/4	13	83	4	23	67	10	–	–	–
Persimmon[a]	2011/10/20	9	91	–	19	81	–	–	–	–
Persimmon	2013/10/8	11	85	4	16	74	11	22	71	7
Persimmon	2014/10/9	11	85	4	16	81	2	19	76	5
Loquat	2011/6/7	22	47	31	34	44	21	–	–	–
Loquat-1[b]	2013/6/19	18	59	22	29	50	21	20	39	41
Loquat-2[b]	2013/6/19	12	69	18	19	69	12	16	50	34
Loquat	2014/6/4	16	62	22	22	56	22	22	43	35
Loquat	2014/6/11	11	68	22	15	72	13	13	48	39
Japanese apricot[c]	2012	12	75	13	12	83	6	–	–	–
Japanese apricot	2014/6/4	17	69	14	11	82	7	22	70	8

[a]Data from [15]
[b]Samples were taken from different trees on the same date
[c]Data from [13]

The results for fruits are listed in Table 18.4. For persimmon and loquat fruits, compared to the relative proportions of skin mass to the total fruits, partitioning percentages of ^{137}Cs and ^{40}K in the skin samples were 1.4–2.2 and 1.2–2.0 times higher, respectively, and those in flesh were 0.8–1.1 and 0.6–0.9, respectively. Thus, ^{137}Cs concentrations in skin were higher than those in flesh for both species. On the other hand, Japanese apricot showed higher percentage of ^{137}Cs distribution than wet mass proportion, although ^{40}K distributions were similar to those of wet mass proportion. In this study, we measured only one sample, however, similar results were obtained in 2012 [13], as listed in the same table.

In 2011, loquat and persimmon fruits were also measured and the ^{137}Cs distribution percentages in skin samples [15] were higher than those observed in 2013 and 2014 for both species. It was assumed that immediately after the ^{137}Cs was taken up through these trees' aboveground parts, it transferred to their growing tissues including the fruits; during development of fruits, fruit flesh mass increases at the middle-late ripening stage [16], but probably ^{137}Cs supply reduced due to the smaller amount of ^{137}Cs uptake through tree surface. By this process, ^{137}Cs distributions in fruit parts differed from those we observed in 2013 and 2014, although more studies are necessary to understand the mechanisms of Cs transfer in fruit trees.

18.4 Conclusions

The effects of washing and boiling of wild edible plants, and peeling of home-grown fruits were studied. It was found that washing plant surface is not effective in removing both ^{40}K and ^{137}Cs, because both elements are in the plant tissues. Food processing retention factor, F_r, decreased by boiling, and the average values ranged from 0.40 to 0.61 for leaf-blade of giant butterbur, mugwort, water dropwort, and fertile stem of field horsetail. Bamboo shoots and petioles of giant butterbur (except soaking in water for 1 h), however, showed slightly higher F_rs of ^{137}Cs of 0.59–0.94, because these tissues are thicker than leaf tissues, and thus, it is reasonable to expect that more radiocesium would be retained in these tissues. For fruits of persimmon and loquat, ^{137}Cs concentrations in skin were higher than those in flesh, but different trend was observed in Japanese apricot. Interestingly, the distributions of ^{137}Cs in skin, flesh, and seeds observed in this study differed from those observed in 2011. The mechanism is not clarified yet; therefore, further studies are necessary to understand the radiocesium transfer to fruits after direct deposition to fruit tree surfaces.

Acknowledgement This work was partially supported by the Agency for Natural Resources and Energy, the Ministry of Economy, Trade, and Industry (METI), Japan.

References

1. Ministry of Health, Labour and Welfare (MHLW) (2015) Monthly report of test results of radionuclide in foods sampled since 01 April 2012 (by date). http://www.mhlw.go.jp/stf/kinkyu/0000045281.html. Accessed 7 Apr 2015
2. Tagami K, Uchida S (2013) Comparison of food processing retention factors of ^{137}Cs and ^{40}K in vegetables. J Radioanal Nucl Chem 295:1627–1634
3. Tagami K, Uchida S, Ishii N (2012) Extractability of radiocesium from processed green tea leaves with hot water: the first emergent tea leaves harvested after the TEPCO's Fukushima Daiichi Nuclear Power Plant accident. J Radioanal Nucl Chem 292:243–247
4. Tagami K, Uchida S (2012) Radiocaesium food processing retention factors for rice with decreasing yield rates due to polishing and washing, and the radiocaesium distribution in rice bran. Radioisotopes 61:223–229
5. Okuda M, Hashiguchi T, Joyo M, Tsukamoto K, Endo M, Matsumaru K, Goto-Yamamoto M, Yamaoka H, Suzuki K, Shimoi H (2013) The transfer of radioactive cesium and potassium from rice to sake. J Biosci Bioeng 116:340–346
6. Goto-Yamamoto N, Koyama K, Tsukamoto K, Kamigakiuchi H, Sumihiro M, Okuca M, Hashiguchi T, Matsumaru K, Sekizawa H, Shimoi H (2014) Transfer of cesium and potassium from grapes to wine. Am J Enol Viticult 65:143–147
7. Nabeshi H, Tsutsumi T, Hachisuka A, Matsuda R (2013) Variation in amount of radioactive cesium before and after cooking dry shiitake and beef. Food Hyg Safe Sci 54:65–70

8. Nabeshi H, Tsutsumi T, Hachisuka A, Matsuda R (2013) Reduction of radioactive cesium content in beef by soaking in seasoning. Food Hyg Safe Sci 54:298–302
9. Nabeshi H, Tsutsumi T, Hachisuka A, Matsuda R (2013) Reduction of radioactive cesium content in pond smelt by cooking. Food Hyg Safe Sci 54:303–308
10. Hachinohe M, Naito S, Akashi H, Todoriki S, Matsukura U, Kawamoto S, Hamamatsu S (2015) Dynamics of radioactive cesium during noodle preparation and cooking of dried Japanese udon noodles. Nippon Shokuhin Kagaku Kogaku Kaishi 62:56–62
11. Sekizawa H, Yamashita S, Tanji K, Okoshi S, Yoshioka K (2013) Reduction of radioactive cesium in apple juice by zeolite. Nippon Shokuhin Kagaku Kogaku Kaishi 60:212–217
12. Sekizawa H, Yamashita S, Tanji K, Yoshioka K (2013) Dynamics of radioactive cesium during fruit processing. Nippon Shokuhin Kagaku Kogaku Kaishi 60:718–722
13. Radioactive Waste Management Funding and Research Center (2013) Removal of radionuclide by food processing – radiocesium data collected in Japan. In: Uchida S (ed) RWMC-TRJ-13001-2, RWMC, Tokyo (in Japanese)
14. International Atomic Energy Agency (2010) Handbook of parameter values for the prediction of radionuclide transfer in terrestrial and freshwater environments. Technical report series no.472, IAEA, Vienna
15. Tagami K, Uchida S (2014) Concentration change of radiocaesium in persimmon leaves and fruits – observation results in 2011 spring −2013 summer. Radioisotopes 63:87–92
16. Arai N (2004) Fruit forms and development. In: Yamazaki K, Kubo Y, Nishio T, Ishihara K (eds) Encyclopedia of agriculture. Yokendo, Tokyo, pp 1122–1123 (in Japanese)

Chapter 19
Monte Carlo Evaluation of Internal Dose and Distribution Imaging Due to Insoluble Radioactive Cs-Bearing Particles of Water Deposited Inside Lungs via Pulmonary Inhalation Using PHITS Code Combined with Voxel Phantom Data

Minoru Sakama, Shinsaku Takeda, Erika Matsumoto, Tomoki Harukuni, Hitoshi Ikushima, Yukihiko Satou, and Keisuke Sueki

Abstract The role of this study in terms of health physics and radiation protection has been implemented to evaluate the internal dose (relative to the committed equivalent dose) and the dose distribution imaging due to gamma rays (photons) and beta particles emitted from the radioactive Cs-bearing particles in atmospheric aerosol dusts deposited in the lungs via pulmonary inhalation. The PHITS code combined with voxel phantom data (DICOM formats) of human lungs was used. We have dealt with the insoluble radioactive Cs-bearing particles of water (about ϕ 2.6 μm diameter) migrated onto any of six regions, ET1, ET2, BB, AI-bb, LNET, and LNTH, in a respiratory system until dropping into blood vessels. Source parameters were those of an adult male breathing a typical air volume outdoors; in the simulated atmosphere (such as systematically setting up a field) those particles would be released on 21:10 March 14 to 9:10 March 15, 2011 in Tukuba, Japan,

M. Sakama (✉)
Department of Radiation Science and Technology, Institute of Biomedical Sciences, Tokushima University Graduate School, Kuramoto-cho 3-18-15, Tokushima, Japan
e-mail: minorusakama@tokushima-u.ac.jp

S. Takeda • E. Matsumoto
Department of Radiation Science and Technology, Institute of Biomedical Sciences, Tokushima University Graduate School, Tokushima, Japan

T. Harukuni • H. Ikushima
Department of Therapeutic Radiology, Institute of Biomedical Sciences, Tokushima University Graduate School, Tokushima, Japan

Y. Satou • K. Sueki
Faculty of Pure and Applied Science, University of Tsukuba, Tsukuba, Japan

© The Author(s) 2016
T. Takahashi (ed.), *Radiological Issues for Fukushima's Revitalized Future*,
DOI 10.1007/978-4-431-55848-4_19

as a filter sampling condition already reported by Adachi et al. In this chapter, we discuss the internal dose and the dose distribution imaging in each voxel phantom for human lung tissues corresponding to the respiratory tracts of BB and AI-bb, respectively.

Keywords Monte Carlo simulation • Committed equivalent dose • Cesium-134 • Cesium-137

19.1 Introduction

At 14:46 JST (Japan Standard Time) March 11th, 2011, a huge earthquake (magnitude $Ms = 9.0$ at the epicenter; epicenter depth is about 10 km) occurred at sanriku off the coast of eastern Japan. A Great tsunami followed the earthquake and caused serious accidents at the Fukushima Daiichi Nuclear Power Plant (FDNPP) and plant workers had been unable to control the cooling system of nuclear fuels at all the plants 1, 2, 3, and 4. At that time, great amounts of hydrogen gases had been produced around their nuclear fuel rods and the remaining fuel assemblies had not cooled off, filling the plants with hydrogen gases; after a while the emergency condition caused hydrogen explosions at plants 1 (on March 12, 2011) and 3 (on March 14, 2011), and fires also occurred at plants 2 and 4 due to those explosions. These serious accidents had caused the release of large amounts of radioactive materials into the environment together with plumes.

A few years after this accident, we had slightly understood some findings regarding all the radioactivity for released radioactive cesium isotopes of Cs-134 and Cs-137 into the environment and the exact migrations for the radioactive plumes including those radioactive materials upon atmospheric conditions [1–5]. Four years elapsed and it has now become clear that the radioactive materials have chemical and physical properties concerning chemical forms, particle sizes, shape, phases (gas or aerosol), water solubility, and residence time [6–10]. Adachi et al. [11] reported that they directly observed spherical Cs-bearing particles emitted during a relatively early stage (March 14–15) of the accident, and also stated that the spherical Cs-bearing particles were larger (their diameters were approximately 2 μm), and they were less water soluble than sulfate particles. In their report, they investigated the coexistence of spherical Cs-bearing particles with Fe, Zn, and possible other elements using SEM and EDS mapping images with the elemental analysis spectrum.

In addition, Satou et al. [12] similarly reported that they investigated whether specific particles such as the spherical Cs-bearing particles observed by Adachi et al. were included in soil samples at the northwestern area about 20 km away from FDNPP where radioactive plumes migrated on March 15, 2011. It was found that their soil samples contained the same spherical Cs-bearing particles indicated by Adachi et al. [11] and their observed particle size and contained elements were about 2 μm diameter with Fe, Zn, and Rb using the same SEM and EDS analyses; they concluded that their obtained properties of these particles in the soil samples were consistent with those reported by Adachi et al. [11]

As some experimental uptakes allow us to understand step by step the chemical and physical properties of the released radioactive Cs-suspending materials taking forms such as spherical Cs-bearing particles into the plumes, we consider that there exists a need to estimate the health effects on the large crowd of people who were outside around FDNPP or away from there when the plumes migrated after the accident. In the course of this research regarding health physics, we have offered some evaluation materials for the health effects based on calculation of the internal dose and the dose distribution imaging due to beta particles and photons emitted from the Cs-bearing particles deposited inside the lungs in pulmonary inhalation using the PHITS ver. 2.76 (Particle and Heavy Ion Transport code System) code [13] combined with voxel phantom data [14] (formed by the DICOM format) of a human lung next to human respiratory tracts.

In the present work, taking source parameters on the calculation includes the radioactive Cs-bearing particles (about ϕ 2.6 μm diameter) distributed onto each of six regions, ET1, ET2, BB, AI-bb, LNET, and LNTH, in a human respiratory tract group until dropping into blood vessels on the assumption of one adult male breathing a typical air volume outdoors during March 14, 21:10 (local time) to March 15, 09:10, 2011 in Tukuba, Japan [11]. We have evaluated the internal exposure and the dose distribution imaging in each of the lung voxel phantoms corresponding to the regions of BB and AI-bb, respectively, using the PHITS code. The period of our interest includes an aerosol sampling filter with a maximum radioactivity level of about $1.00E+04$ mBq/m^3 from the radioactive plume and also the radioactive Cs-bearing spherical particles found in the filter, as reported by Adachi et al. [11].

19.2 Method and Materials

Considering internal exposure results from intake of an unsealed radioactive material, in this case treated as the radioactive Cs-bearing particles of Cs-134 and Cs-137, to the human body, it seems natural that there are three pathways into the body for the radioactive material to take:

1. Inhalation: entry through organs in a respiratory tract
2. Ingestion: entry through gastrointestinal tract
3. Direct absorption: absorption through intact or wound

It should be noted that the inhalation pathway is most closely connected with a suspending material such as an aerosol cluster in the atmosphere. In the human body, the respiratory tract is mainly in charge of inhalation. To estimate internal exposure at each of the organs or tissues using all calculation methods such as Monte Carlo simulation codes, there is a need for appropriate mathematical models to describe the various processes involved in the internal deposition and retention of the cesium nuclides and the associated radiation doses of gamma rays (photons) and beta particles emitted from these decays received by various organs and tissues in the respiratory tract. As a general standard model, the respiratory tract is viewed as a series of compartments into which the cesium-bearing particles enter and exit at various rates, ultimately being removed from the respiratory tract regions through exhalation and mechanical clearance processes to outside the body, gastrointestinal tract, and blood, respectively. We have stated here that a deposited dose at each organ or tissue, which is due to various radiation, should really be equivalent to a "committed equivalent dose" including the well-known one of radiation safety terms. That is, in this study, the committed equivalent dose should be estimated extremely at each compartment composed of respiratory tract regions using the PHITS code based on a Monte Carlo calculation method.

Following the recommendations by the ICRP in publication 60 [15] and the model of the human respiratory tract in publication 66 [16], we had properly specified PHITS calculation parameters in input files on which geometrical definitions of PHITS's own code (so-called PHITS language) were interpreted as compartment models of the human respiratory tracts. This compartment modeling was conformed for children aged 3 months, 1, 5, and 10 years and for male and female 15-year-olds and adults of a general population stated on the ICRP publications. The respiratory tract regions were represented by five regions: the Extrathoracic (ET) airways were divided into ET1, the anterior nasal passage, and ET2, consisting of the posterior nasal and oral passages, the pharynx and larynx. The thoracic regions were Bronchial (BB), Bronchiolar (bb), and Alveolar-Interstitial (AI, the gas exchange region). For the bb and AI, inasmuch as both are located next to each other around respiratory terminal bronchioles, in this study we regarded them as a unity region of AI+bb. Lymphatics were associated with the extrathoracic and thoracic airways (LNET and LNTH). Their regions were defined based principally on radiobiological considerations, but also taking account of differences in respiratory function, deposition, and clearance for the generation member. Each PHITS calculation at those six regional compartments, ET1, ET2, BB, AI-bb, LNET, and LNTH, was implemented by 100,000 histories of photon (gamma ray) and beta particle (e^-) decaying Cs-134 and Cs-137 nuclides, respectively, with three different types, type-F (fast), type-M (middle), and type-S (slow), of cesium transfer coefficients into a respiratory tract in a body, assuming that the radioactive Cs-bearing particles ($2.6\,\mu$m diameter) were deposited at each tissue of those six tissue region compartments. The transfer coefficients were obtained by the WinAct package software [17] based on the ICRP publication 66. This package software is the ORNL numerical solver (Windows version) for the coupled set of differential equations describing the kinetics of a radionuclide in the body.

With the present PHITS calculation method of the committed equivalent doses due to the Cs-bearing particles on the respiratory tract regions, it is very important to evaluate U(50) and SEE (Specific Effective Energy) values. U(50) gives the total number of decaying nuclides which depends on the decay property and the elemental transfer coefficient into a tissue of interest and the value has been evaluated using the WinAct software. Setting three types of differences in transfer coefficients from type-F to type-M and type-S, we have estimated the committed equivalent doses upon the three cesium transfer conditions on each type of U(50). SEE values have the relationship among radiation branching ratios, Y, radiation energies for gamma rays and average beta particles of Cs-134 and Cs-137, E, specific absorbed fraction, SAF, and radiation weighting factor, w, and this value is given in the function:

$$SEE = Y \times E \times SAF \times w. \tag{19.1}$$

The general standard method includes the general committed equivalent dose that thus far has been estimated using the SEE value based on the empirically recommended SAF value of ICRP publication 66. Using the present PHITS calculation method, its code has provided the committed equivalent dose on which the PHITS output data directly result in a relative specific absorbed dose per a radiation source event (Gy/source). Therefore, as it is strongly confirmed that it should correspond to the fraction of SAF as shown in Eq. (19.1), the SEE value can be given as involving the SAF value, and we are consequently able to provide the committed equivalent doses (Sv/particles) for each respiratory tract using the SEE involved on the calculated SAF value, Y, and w in Eq. (19.1).

This PHITS code simulates initial beta and gamma decay properties [18] emitted from Cs-134 and Cs-137 nuclides into the Cs-bearing particles and secondary radiation associated their radiation event by event in a virtualized space. The input files include voxel phantom data compiled for PHITS code into which it has given geometrical and material information of the respiratory tracts and source information of the initial radiations generated uniformly at each respiratory tract of interest as a source region. In other words, the committed equivalent dose has also been provided with the summation of absorbed doses on a lung tissue event by event into a compartment respiratory tract model by adding U(50) and radioactivity ratio (a ratio of about one to one; this is experimentally well-known from many reports of investigations of specific radioactivity for various environmental samples) of Cs-134 and Cs-137 at an early stage of this nuclear accident to the PHITS input data.

For the present treatment of voxel computational phantoms referred to the adult reference male of ICRP 110 publication [19] we have compiled its CT values [20] formed by the DICOM format to the PHITS input data using the "dicom2phits" package program combined with the PHITS calculation code. This program has involved the data exchange functions based on the relationship between CT values [19, 20] and material densities and compositions at each tissue or organ [14]. In the present PHITS input file, we have generated the voxel phantom data with the smallest unit size of $0.98 \times 0.98 \times 10 \, mm^3$ for each compartment of the respiratory tract regions around a human lung.

19.3   Results

19.3.1   U(50): Total Number of Decaying Nuclides at Respiratory Tracts of BB and AI

Table 19.1 summarized the total numbers U(50) of decaying Cs-134 and Cs-137 nuclides at respiratory tracts of BB and AI in comparison with their three different types, type-F (fast), type-M (middle), and type-S (slow). Both the BB and AI tissues shown in Table 19.1, indicate that their U(50) values of type-S (small) for Cs-134 and Cs-137 should become obviously highest with a decreasing cesium transfer coefficient so that the cesium elements have been deposited at each respiratory tract for a long time. In particular, there is a very remarkable difference among the type-F, type-M, and type-S at the AI respiratory tract due to the very large surface area of small alveoli around respiratory bronchioles.

19.3.2   Internal Exposure Dose Distribution Imaging for a Respiratory Tract of BB

Figure 19.1a shows the whole systematic upper half of the human body constructed from the voxel phantom data compiled by the PHITS language. The geometrical and material voxel phantom data around a lung are described based on phantom data of an adult male with CT value using the "dicom2phits" package program combined with the PHITS calculation code. The other figures illustrate the calculated internal exposure dose distributions (expressed at the vertical axis of the absorbed dose (Gy) at T-heat mode on PHITS) of type-S in Fig. 19.1b and type-F in Fig. 19.1c for the BB respiratory tract of the present interest. Their absorbed doses expressed into the figures contribute to whole radiation energies absorbed at each tissue due to the primary gamma rays emitted from Cs-134 and Cs-137 decaying nuclides at the BB respiratory tract. The figures represent that both distributed intensities of absorbed dose for type-S and type-F have increased as formed outlines of the compartment boundaries of BB respiratory tract because of the radiation source generation in there. In addition, in comparison between the figures, it has been found that both generated radiation has been gradually transported on the same order of about 10^3–10^4 as the intensity distributions across the entire upper half of the body. And

Table 19.1 The total numbers of decaying Cs-134 and Cs-137 nuclides at respiratory tracts of BB and AI

Tissue	Cs-134			Cs-137		
	Type-F	Type-M	Type-S	Type-F	Type-M	Type-S
AI	275.17	2843600	10958000	271.85	3186500	23961000
BB	22.936	57487	73884	22.659	58184	75069

Fig. 19.1 (**a**) Shows the entire systematic upper half of the human body constructed from the voxel phantom data compiled by PHITS language. (**b**, **c**) Illustrate the calculated internal exposure dose distributions expressed at the vertical axis of absorbed dose (Gy) obtained from T-heat results on PHITS

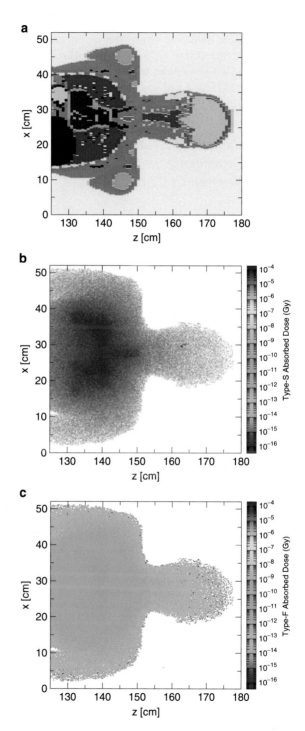

also the absorbed dose distribution of type-S was much higher than that of type-F, because it is definitely likely that it was caused by the total number U(50) of type-S more than that of type-F into the BB region due to too many extremely late Cs-134 and Cs-137 nuclides transferring in the region upon the transfer coefficients of type-S to the blood and gastrointestinal tract.

19.3.3 Committed Equivalent Doses in Comparison with Dependence on Gamma Rays and Beta Particles Between Cs-134 and Cs-137

Tables 19.2 and 19.3 show calculated committed equivalent doses in a comparison with dependence on gamma rays and beta particles associated with the decaying nuclides of Cs-134 and Cs-137 between two methods of PHITS and the general standard method, and also among transfer coefficients of type-F, type-M, and type-S, respectively. In this chapter, we have focused on the calculated result data in a comparison between BB and AI+bb respiratory regions. Comparing all the factors of radiation, transfer coefficient, and respiratory tract in our obtained data as shown in Tables 19.2 and 19.3, it is very likely that the present PHITS simulated method can reproduce those committed equivalent doses estimated approximately by the general standard method. Especially, it is really remarkable that the committed equivalent

Table 19.2 Comparison with calculated committed equivalent doses of Bronchial (BB) and a unity region of Alveolar-Interstitial (AI) and Bronchiolar (bb), which are associated with gamma rays of Cs-134 and Cs-137, by PHITS and general standard method

		Cs-134: gamma ray (Sv/particles)			Cs-137: gamma ray (Sv/particles)		
		Type-F	Type-M	Type-S	Type-F	Type-M	Type-S
PHITS	BB	2.85602E-11	4.26206E-08	1.57344E-07	1.60146E-11	1.86545E-08	1.43411E-07
	AI+bb	2.56911E-11	3.74353E-08	1.41104E-07	1.30911E-11	1.71161E-08	1.29311E-07
General standard method	BB	3.34002E-11	3.95500E-08	1.45290E-07	1.32564E-11	1.78343E-08	1.36558E-07
	AI+bb	2.73125E-11	3.76565E-08	1.41924E-07	1.08211E-11	1.70816E-08	1.28834E-07

Table 19.3 Comparison with calculated committed equivalent doses of Bronchial (BB) and a unity region of Alveolar-Interstitial (AI) and Bronchiolar (bb), associated with beta particles of Cs-134 and Cs-137, by PHITS and general standard method

		Cs-134: β-particle (Sv/particles)			Cs-137: β-particle (Sv/particles)		
		Type-F	Type-M	Type-S	Type-F	Type-M	Type-S
PHITS	BB	1.62220E-11	2.18575E-08	4.03264E-08	1.71321E-11	1.06724E-06	7.89293E-06
	AI+bb	7.15408E-12	6.32775E-08	2.40142E-07	8.40650E-12	8.73540E-08	6.46041E-07
General standard method	BB	4.21726E-10	4.60743E-07	6.01231E-07	5.86399E-10	6.60018E-07	9.03500E-07
	AI+bb	1.94080E-11	7.51489E-08	2.45339E-07	2.66325E-11	1.02437E-07	6.11277E-07

doses attributed to the gamma rays of Cs-134 and Cs-137 on the PHITS calculation results are entirely consistent with those of the general standard method at both the respiratory tracts of BB and AI+bb.

19.4   Discussion

In the present study, providing that the radioactive Cs-bearing particles around 21:00 March 14 to 9:00 March 15, 2011, which were probably released from the FDNPP at the early stage of the nuclear accident, are supposed to have been deposited into human lung tissues from the main to tip regions corresponding to BB and AI+bb respiratory tracts, we have discussed that the PHITS method [13] would probably be available for a Monte Carlo evaluation of committed equivalent dose and the relative internal dose distribution imaging based on possible systematic and optional arrangements using the human voxel phantom by itself or a nearby generation, for example, Japanese adult Male (JM) voxel phantom [14], for anyone with health concerns in Fukushima.

Through a course of the present PHITS Monte Carlo simulation method that has calculated the deduced committed equivalent doses for several respiratory tracts using voxel phantom data of adult human lung tissues and the absorbed dose distribution imaging due to gamma rays and beta particles, the Cs-bearing particles, we have confirmed that the calculated committed equivalent doses contributed to the gamma rays of Cs-134 and Cs-137 can fairly reproduce these results led by the general standard method as shown in Table 19.2. From the absorbed dose distribution in Fig. 19.1b, considering that it is very likely that the Cs-bearing particles with less water solubility would belong to the transfer coefficient for Type-S, it should be noted that transmitted gamma rays from the BB respiratory tract influence part of the radiation energies on surrounding tissues with a decreasing two to three orders of absorbed dose magnitude of BB itself per voxel into the upper half of the body. In regard to beta particles attributed to Cs-134 and Cs-137 in the Cs-bearing particles at both the BB and AI+bb respiratory tracts, as shown in Table 19.3, those committed equivalent doses evaluated by PHITS are inconsistent with those calculated from general standard method for any type, in particular for Type-S, in comparison with the case of the gamma rays. The difference can be explained by the incorrect exchange of all the actual tissue sizes into AI+bb regions to PHITS input data because CT values in the present DICOM data have a minimum-limited unit size for shaping information of these tracts. And thus, the minimum-limited size would be larger than the actual tissue size and it is very likely that there are wide opening spaces between each of the tissues into the AI+bb regions. We have, therefore, deduced that the present PHITS calculation could not reproduce the transport simulations of beta particles for Cs-134 and Cs-137 due to charged beta particles whose property is definitely involved in the stopping power higher than that of non charged gamma rays, and thus the beta particle may be attributed to be greatly influenced by the difference of each opening space size around AI+bb. Then, the rate of transmitting beta particles would increase without these energy losses at AI+bb regions where they should lose the radiation energies themselves.

From the viewpoint of Monte Carlo evaluation of the internal dose and the distribution imaging due to the insoluble radioactive Cs-bearing particles of water deposited in the lung tissues of a human (adult male) via pulmonary inhalation around 21:00 March 14 to 9:00 March 15, 2011, this work will bear out that the PHITS method was able to evaluate that the committed equivalent dose of BB is 1.57344E-07 (Sv/particles) and 4.03264E-08 (Sv/particles) associated with gamma rays and beta particles of Cs-134, 1.43411E-07 (Sv/particles), and 7.89293E-06 (Sv/particles) associated with gamma rays and beta particles of Cs-137, respectively, at Type-S reproducing the migration next to an actual transfer coefficient into lung tissues of the simulated adult male person. Additionally, we have found that the PHITS method is primarily available not only for evaluating the distribution of internal doses of any tissues of interest relative to both the absorbed doses and the committed equivalent doses, but also investigating their doses at any small piece such as a voxel level depending on CT value and taking the gamma ray and beta particle specific spectra to elucidate the dose distribution mechanism in detail; we also have stated that there is a need to improve the time-consuming and statistical precision contributing to more correct calculations using the PHITS code in the future.

19.5 Conclusion

We have shown that the present PHITS code, which is combined with voxel phantom data of human lung tissues for an adult male, can approximately evaluate the internal dose relative to the committed equivalent dose and the distribution imaging due to gamma rays and beta particles associated with decaying Cs-134 and Cs-137 at the insoluble radioactive Cs-bearing spherical particles of water probably released during 21:10 March 14 to 9:10 March 15, 2011 in Tukuba, Japan, as reported by Adachi et al. In particular, it was found that the calculated committed equivalent doses due to the gamma-rays of Cs-134 and Cs-137 were entirely consistent with those of the general standard method at both the respiratory tracts of BB and AI+bb using the SEE value based on ICRP publication 66, and the estimated committed equivalent doses were 1.57344E-07 (Sv/particles) and 1.41104E-07 (Sv/particles) for Cs-134 and 1.43411E-07 (Sv/particles) and 1.29311E-07 (Sv/particles) for Cs-137 on both Type-S conditions such as a specific transfer coefficient corresponding to the migration rate of cesium until dropping into blood vessels around lung tissues. The PHITS code will probably be available for evaluation of committed equivalent dose and the relative internal dose distribution imaging based on possible systematic and optional arrangements using the human voxel phantom by itself or nearby generation for anyone with health concerns in Fukushima. We plan in the near future to improve several issues of the time-consuming and statistical errors in order to implement more correct calculations by the PHITS method and to extend this method to voxel phantom data of the entire human body including various public fields in the environment using the PHITS code coupled with MCAM and RVIS [21].

Acknowledgements The authors are grateful to research group members for radiation transport analysis in Japan Atomic Energy Agency (JAEA), Dr. K. Manabu, Dr. K. Satou, Dr. S. Hashimoto, Dr. T. Furuta, and Dr. T. Satou, for operating voxel phantom data in extremely large file sizes for an adult male and also for implementing the PHITS calculations again and again to make an attempt to optimize the calculation method.

References

1. Manolopoulou M, Vagena E, Stoulos S, Ioannidou A, Papastefanou C (2011) Radioiodine and radiocesium in Thessaloniki, Northern Greece due to the Fukushima nuclear accident. J Environ Radioact 102:796–797
2. Amano H, Akiyama M, Chunlei B, Kawamura T, Kishimoto T, Kuroda T, Muroi T, Odaira T, Ohta Y, Takeda K, Watanabe Y, Morimoto T (2012) Radiation measurements in the Chiba Metropolitan area and radiological aspects of fallout from the Fukushima Dai-ichi nuclear power plants accident. J Environ Radioact 111:42–52
3. Momoshima N, Sugihara S, Ichikawa R, Yokoyama H (2012) Atmospheric radionuclides transported to Fukuoka, Japan remote from the Fukushima Dai-ichi nuclear power complex following the nuclear accident. J Environ Radioact 111:28
4. Katata G, Chino M, Kobayashi T, Terada H, Ota M, Nagai H, Kajino M, Draxler R, Hort MC, Malo A, Torii T, Sanada Y (2015) Detailed source term estimation of the atmospheric release for the Fukushima Daiichi nuclear power station accident by coupling simulations of an atmospheric dispersion model with an improved deposition scheme and oceanic dispersion model. Atmos Chem Phys 15:1029–1070
5. Haba H, Kanaya J, Mukai H, Kambara T, Kase M (2012) One-year monitoring of airborne radionuclides in Wako, Japan, after the Fukushima Dai-ichi nuclear power plant accident in 2011. Geochem J 46:271–278
6. Doi T, Masumoto K, Toyoda A, Tanaka A, Shibata Y, Hirose K (2013) Anthropogenic radionuclides in the atmosphere observed at Tsukuba: characteristics of the radionuclides derived from Fukushima. J Environ Radioact 122:55–62
7. Miyamoto Y, Yasuda K, Magara M (2014) Size distribution of radioactive particles collected at Tokai, Japan 6 days after the nuclear accident. J Environ Radioact 132:1–7
8. Kaneyasu N, Ohashi H, Suzuki F, Okuda T, Ikemori F (2012) Sulfate Aerosol as a potential transport medium of radiocesium from the Fukushima nuclear accident. Environ Sci Technol 46:5720–5726
9. Niimura N, Kikuchi K, Tuyen ND, Komatsuzaki M, Motohashi Y (2015) Physical properties, structure, and shape of radioactive Cs from the Fukushima Daiichi nuclear power plant accident derived from soil, bamboo and shiitake mushroom measurements. J Environ Radioact 139:234–239
10. Yanaga M, Oishi A (2015) Decontamination of radioactive cesium in soil. J Radioanal Nucl Chem 303:1301–1304
11. Adachi K, Kajino M, Zaizen Y, Igarashi Y (2013) Emission of spherical cesium-bearing particles from an early stage of the Fukushima nuclear accident. Sci Rep 3:2554:1–5

12. Satou Y, Sueki K, Sasa K, Kitagawa J, Ikarashi S, Kinoshita N (2015) Vertical distribution and formation analysis of the ^{131}I, ^{137}Cs, ^{129m}Te, and ^{110m}Ag from the Fukushima Dai-ichi nuclear power plant in the beach soil. J Radioanal Nucl Chem 303:1197–1200
13. Sato T, Niita K, Matsuda N, Hashimoto S, Iwamoto Y, Noda S, Ogawa T, Iwase H, Nakashima H, Fukahori T, Okumura K, Kai T, Chiba S, Furuta T, Sihver L (2013) Particle and heavy ion transport code system, PHITS, version 2.52. J Nucl Sci Technol 50:913–923
14. Sato K, Noguchi H, Emoto Y, Koga S, Saito K (2007) Japanese adult male voxel phantom constructed on the basis of CT images. Radiat Prot Dosimetry 123(3):337–344
15. International Commission on Radiological Protection (1991) 1990 recommendations of the international commission on radiological protection. Publication 60. Annals of the ICRP, vol 21/1–3. Pergamon Press, Oxford
16. International Commission on Radiological Protection (1994) Human respiratory tract model for radiological protection. Publication 66. Annals of the ICRP, vol 24/1–3. Pergamon Press, Oxford
17. Leggett RW et al (1993) An elementary method for implementing complex biokinetic models. Health Phys 64:260–278
18. Firestone RB, Shirley VS (eds) (1996) Table of isotopes, 8th edn. Wiley, New York
19. International Commission on Radiological Protection (2009) 1990 adult reference computational phantoms. Publication 110. Annals of the ICRP, vol 39/2. Pergamon Press, Oxford
20. Schneider W (2000) Correlation between CT numbers and tissue parameters needed for Monte Carlo simulations of clinical dose distributions. Phys Med Biol 45:459–478
21. Wu Y, FDS Team (2009) CAD-based interface programs for fusion neutron transport simulation. Fusion Eng Des 84:1987–1992

Chapter 20
A Study of a Development of Internal Exposure Management Tool Suited for Japanese Diet Behavior

Shin Hasegawa, Shinya Oku, Daisuke Fujise, Yuki Yoshida, Kazuaki Yajima, Yasuo Okuda, Thierry Schneider, Jacques Lochard, Isao Kawaguchi, Osamu Kurihara, Masaki Matsumoto, Tatsuo Aono, Katsuhiko Ogasawara, Shinji Yoshinaga, and Satoshi Yoshida

Abstract After the Fukushima Nuclear Power Plant accident, one of the main issues at stake was the potential intake of contaminated foodstuff by residents of the affected areas. In this context, the importance of the management of the internal exposure by food intake has emerged. For this purpose, a system was developed for estimating the amount of the radioactivity ingested through the diet in order to manage the internal exposure evolution of exposed people. The ultimate goal of this system is to consider all the radiation exposure data including medical exposure in an integrated manner.

In this perspective, a tool that was used for internal exposure assessment in Europe after the Chernobyl disaster has been adapted to be suitable to Japanese diet behavior. The tool was implemented in a Web application in order to estimate the amount of radioactivity in the dish and to manage the internal exposure history of the individuals. This system automatically collects the test results of radionuclide in foods available on the web.

It manages the individual internal exposure history estimating the amount of radioactivity in the ingested dish. The developed application enables individual to manage his/her protection by checking radioactivity ingestion history and

S. Hasegawa (✉) • D. Fujise • Y. Yoshida • K. Yajima • Y. Okuda • I. Kawaguchi • O. Kurihara •
M. Matsumoto • T. Aono • S. Yoshinaga • S. Yoshida
National Institute of Radiological Sciences, 4-9-1, Anagawa, Inage-ku, Chiba City,
Chiba Prefecture, Japan
e-mail: haseshin@nirs.go.jp

S. Oku
Regle Co. Ltd., Fukuro-machi, Shinjuku-ku, Tokyo, Japan

T. Schneider • J. Lochard
Nuclear Evaluation Protection Centre (CEPN), 28 Rue de la Redoute,
92260 Fontenay-aux-Roses, France

K. Ogasawara
Faculty of Health Sciences, Hokkaido University, N12-W5, Kitaku, Sapporo, Japan

© The Author(s) 2016   221
T. Takahashi (ed.), *Radiological Issues for Fukushima's Revitalized Future*,
DOI 10.1007/978-4-431-55848-4_20

determining to eat or not the dish according to the amount of radioactivity in the dish. This system, which has the potential to contribute to the radiation protection culture of people living in the contaminated areas of Fukushima Prefecture, has been evaluated by specialists of radiation protection. The following step will be the test of the system by the individuals themselves.

Keywords Internal exposure management • Internal exposure from food intake • Radioactivity ingestion history management • Radiation protection culture

20.1   Introduction

A large amount of radioactive materials was released into the environment by the Fukushima Daiichi nuclear power plant accident caused by the Great East Japan Earthquake, leading to an increasing concern of the citizens about radiation. As a result, not only the concern about the external radiation exposure from the environment but also the growing concern about internal radiation exposure related to food deepened awareness of the importance of a global exposure dose management. It is the purpose of this research to handle and to unify the information about the radiation exposure for affected individuals.

According to ICRP Publication 111 [1], optimization of radiation protection for the residents living in the contaminated areas is the driving principle. This cannot be achieved without engaging the affected people in the process and in this perspective information and dialogue meetings on radiation protection, and a variety of lectures have been carried out [2]. Upon preliminary investigation of the present study, the authors participated in dialogue meetings, which took place in the affected areas. This was an opportunity to hear the voice of the residents. What we learned was that many people in the absence of measurements related to their own situation were anxious about the situation and as a result had no other choice than imposing themselves restrictions in their diet. This was perceived as a strong constraint on the day-to-day life and induced changes in the cultural habits on food consumption. Although the first measurements made in the Prefecture were indicating concentration values in foodstuff much lower than those initially expected or announced by different voices, the consumers were generally still reluctant to buy products from Fukushima Prefecture. This confusing situation led us to promote a database providing the available information in order to better inform the residents.

For food on the market, radioactivity measurement results have been published by the Ministry of Health, Labour and Welfare (MHLW) [3], local governments [4, 5], and Japan Agricultural Cooperatives (JA) [6]. As for the foods that residents gathered on kitchen gardens, forests, rivers, and in the sea, the community centers have provided equipment to measure the radiation, etc., so it has become possible to know the amount of radioactivity in food in almost all cases. However, the amount of radioactivity is expressed in Bq/kg, and it is difficult for the local residents to directly estimate the influence of these values on the total dose they receive (expressed in mSv). In order to determine themselves whether or not to eat the various food products, it is useful to know how much radioactivity they take when they eat those products, the degree of internal exposure, and also the quantity of radioactivity discharged from the body. In this perspective a tool which provides information about all these aspects was considered as very helpful. After the Chernobyl accident, the ETHOS project and the CORE program [7, 8] in the contaminated areas of Belarus provided education for local residents about radiation protection to favor their involvement in their own protection. A software for internal exposure assessment called CORPORE has been developed and used in this program in cooperation with Norwegian radiation protection experts. With this software, it is possible to describe and analyze the results of whole body counters (WBC) measurements. Using this tool allows registering the amount of radioactivity taken during the daily diet and also allows health care workers to open a dialogue with individuals about the radiological quality of the food. By doing so the tool allows to enhance the awareness of residents about the risk of internal exposure and also to help individuals to make informed choices about their diet. Although using this tool to find the cause in the WBC measurement results is a so-called reverse direction, it allows to know how much radioactivity is consumed. The estimate intake of exposure amount taken when people ingest the food is displayed in a graph, which also shows the personal history of the accumulation and discharge of radiation. By using this graph people can decide what is not suitable to eat/suitable to eat sometimes/the amount suitable to eat/occasionally suitable to eat. Finally this approach allows residents to easily be able to determine what to eat.

In this study by using the mechanisms and models of CORPORE taking into consideration the situation in Japan and Japanese eating habits, we implemented and verified our method of internal exposure management and acquiring information about food.

20.2 Methods

The original CORPORE was implemented to respond to the Belarus and Norwegian situation by handling the food products separately and considering two modes of ingestion: chronic intake and episodic intake. Therefore, its use required to be adapted to the typical Japanese environment. Japanese seldom keep eating the same menu everyday. They often eat various dishes using various food products. It is more natural and easy to manage the ingredients rather than dishes for them. Thus, we decided to handle ingestion as dishes for the new system.

Original CORPORE handles only cesium-137 (Cs-137). Other radionuclides were also required to consider in Japan. It was issued for iodine-131 (I-131), cesium-134 (Cs-134), and Cs-137 radioactivity measurements. Although without any details about each food product, the results were also presented for strontium-90(Sr-90) and plutonium measurement on market basket methods [3]. For this reason, it was decided that the support of multi-nuclide is also required. Original CORPORE was

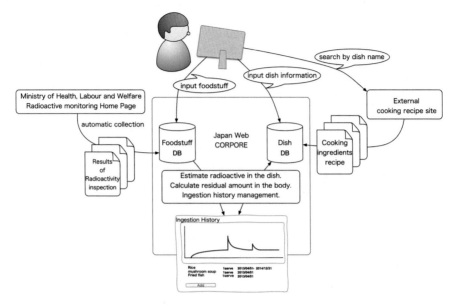

Fig. 20.1 A diagram of Japan Web CORPORE

basically intended for use by radiation protection and medical personnel. In order to be used by residents, the user management system and intake history management system have been developed. Original CORPORE also has statistical functions about the intake status for assessment in order to display the position of the resident in the group. All these features have been kept.

In order to adapt the original CORPORE to the Japanese situation, it is developed as a Web application for widespread use. A database structure allows managing ingestion of food products as dish. It hence supports to various nuclide types. Applying user management function, users can manage their own intake history respectively. To facilitate complicated food information input, food and dish input system is applied so that forward-direction usage can be realized. All these functions are presented in Fig. 20.1 and details are explained below.

20.2.1   System

Local residents can enter their own dish to evaluate their contamination level, so they can make the exposure management of themselves by browsing the radioactivity intake information and internal exposure information. They are also able to access it and use it from personal computers or smart phones through the Internet, as it is developed as a Web application. To operate in typical server configurations, MySQL 5.5 database system, most popular open source database system in the world [9], was used, and it was implemented in PHP 5.5, Hypertext Preprocessor which is a server-side scripting language [10], to work on typical web server applications such as Apache 2.4 [11], etc. Also, it is available from the general viewing environment, and is successfully displayed in Internet Explorer 9 or higher, Google Chrome version 10.0.648 or higher and Safari version 5.0 or higher.

20.2.2   Database Structure

In order to input food intake situations to match the Japanese food environment, we considered the modification of the food database structure of the original CORPORE. It was initially limited to the food database, and we enlarged it to the ingredients database.

Also, the original CORPORE allowed only Cs-137 input, while in the modified version the database structure corresponds to the multi-radionuclide species. Presented in the survey results by the MHLW released I-131, Cs-134, Sr-90 and plutonium (Pu-238, Pu-239, Pu-240) became a possible input.

20.2.3   Food and Cooking Information Input

Radioactivity test results announced from time to time by the MHLW turn into DB automatically, and this information is available to use for products. As for the food which is ingested daily, by applying the technology developed by Kawashima et al. [12], in cooperation with external recipe information sites, it is possible to avoid the trouble of inputting all the information about materials. Cooking information can be reused as a template.

20.2.4   User Management

To perform user management, the user is enabled to browse his/her intakes in the history of radiation continually and record meals.

In addition, it allows managing the users' rights, such as general user, user with administrative privileges; medical user to make the assessment and research user to make the research were prepared.

The medical user privileges include a privilege to view all the information of the population to perform the individual exposure assessment. The research user privileges include a privilege to view anonymized information about the intake of the population.

20.2.5 Form of Use

Original CORPORE was used in Belarus as an assessment tool (1) to determine the amount of radioactivity in the body using the WBC measurements as inputs, (2) to engage a dialogue with the residents on their diet in case the internal contamination levels were significantly higher than the average level, (3) to analyze the estimated exposure dose using the graph to display it; thus, it is the reverse direction to explore the cause from the result. Displaying the radiation dose that is expected by ingesting the food is not only the opposite direction to determine whether to eat or not. Also, the ingested radioactivity substance amount is recorded, because it is assumed that a possible use of its dose can happen in the future again. This data is saved in history of the daily diet, and it can be displayed in a graph to see the remaining radioactivity quantity in the body.

Moreover, it can be used as an assessment tool, since it matches the functions of the original CORPORE, and in the present system the input and statistics of WBC measurement results are displayed, too.

20.3 Results

In the Web version of CORPORE Japan, it is possible to access the Web application through the Internet, perform user management on the basis of the food ingredients input to make a meal, learn about ingested food information for each user, and display a graph of radioactivity intake and biological half-life discharge status, as well as to be able to confirm oneself the graph of intake and discharge status of radioactivity on the browser. From 0-year-old to 1-year-old, from 2-years-old to 5-years-old, from 6-years-old to 10-years-old, from 11-years-old to 15-years-old the ingested radioactivity is reduced by each of the biological half-life factor by 16-years-old of age or older, and the user's age appearance is displayed as a graph (Fig. 20.2).

As an assessment function, similar to the original CORPORE, it is possible to confirm the statistics display about any users' position among the intake average and make groups of the population on the browser (Fig. 20.3).

Furthermore, the product database is updated periodically through the access to the MHLW website which is obtained automatically, to see the results of the radioactivity tests. In April 2015, about 1.2 million cases of food radioactivity test results were published and are available as contamination information. Although it was confirmed that the database allows entering not only Cs-137 information but also other radionuclides, however, currently this system implemented only Cs-137 biological model, hence the result graphs of such other nuclides are not shown.

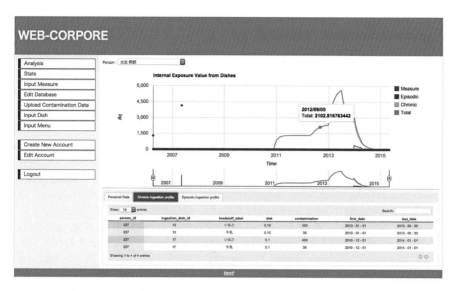

Fig. 20.2 An analysis screen shows the history of radioactivity ingestion and reduction

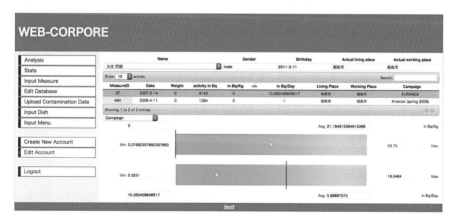

Fig. 20.3 A stats screen shows user's position among the intake average and make groups of the population

20.4 Discussion

20.4.1 Verification of the Usability

As a method of using this system, before cooking any meal or buying any ingredients, one can enter the system from a personal computer or a smart phone, enter the meal ingredients one has access to. By doing this, one will estimate the intake. This way allows obtaining the history of radioactivity intake and estimates the information about the intake of internal exposure dose, so one can make the decision to reduce the amount to eat/not to eat something if it is not suitable/eat everyday or occasionally/not eat at all/so from that previously ingested data the user can decide what and how much to eat or not. When the user ate it, the information is kept in history and becomes the basic judgment criteria for the meals eaten after. Thus, this system can be considered to contribute to the "recovery of the right to self-determination" and is expected to help residents to regain confidence by managing their day-to-day intakes.

Then, as the present system has been constructed keeping in mind the assessment capabilities of the original CORPORE, it is also applicable for reverse utilized assessment. Therefore, it can be used not only in Japan. In fact, we asked the original CORPORE development team to trial this system, and it was highly evaluated. It was also announced to be easier to use than the "original CORPORE because it is only necessary to input each dish, and not each food ingredient as originally and this is obviously more realistic for the Japanese context," "Because ETHOS and CORE activities in Belarus started 10 years after the accident, Cs-134 was ignored in CORPORE given its very small influence at the start of the project. In Japan, the accident has just happened so the knowledge about short nuclides in their half-life corresponded to the multi-nuclides is extremely valuable."

Also, it became possible to handle the natural origins of radioactivity information from the previous self, and to perform the management of global radioactivity uptake.

The number of food products is 1,199,241, which is integrated in the product database. The levels of 1,113,931 products of them (92.8 %) are under the detection limit value. Among them the number of food products added from January to March 2015 is 67,401, among them 65,117 (96.6 %) are about the detection limit value or less, equal to or greater than the limit value (Table 20.1) of the food makes 145 (0.2 %) of cases. For the 145 ingredients of the limit value or more see Table 20.2. The percentage of foods with the limit value or more is very little, mostly it is wild boar meat, but because black bear meat and Japanese deer meat is not common in a general household it is hardly applicable as a food ingredient for Japan. It can be said that the majority of Japanese people at the moment are hardly affected by this food. However, easy evaluation tools are required in contaminated areas for evaluation and the understanding of the situation. These tools will help to develop the radiation protection culture to individuals allowing them to have a grip on their own protection.

Table 20.1 The reference levels of foods in Japan (Notice No. 0315 Articles 1 of the Department of Food Safety, March 15, 2012)

Category	Limit level (Bq/kg)
Water	10
Soft drinks	10
Tea for drinking	10
Milk	50
Foods other than above	100

Table 20.2 A breakdown of the products positive at levels exceeding limits between January 1st and March 31st 2015 in the product database

Name of product	Number of products positive at levels exceeding limits
Wild boar meat	116
Japanese deer meat	10
Asian black bear meat	5
Copper pheasant meat	3
Japanese stingfish	3
Iwana mountain trout	5
Landlocked masu salmon	1
Mushroom powder	1
Stone flounder	1

Moreover, these tools are urgently required in this situation, and it is essential to prepare and keep them in order in case of future contingencies.

20.4.2 Subject of Future Investigation

The system is made to favor the development of the radiation protection culture of people in contaminated areas and its neighborhoods; it will be one of the tools of reconstruction of their living environment, so it is necessary to consider this system for the future.

This system estimates the amount (Bq) of radioactivity that was ingested by using the mechanism of CORPORE, visualizes the state of users' history, and helps to show the discharged radioactivity as a graph. However, the residents of contaminated areas want to know how much of the exposure dose (mSv) they receive and the ingestion part of it. In the future, it will be necessary to establish a mechanism for displaying the conversion of the level of contamination (Bq) into the internal dose (mSv) by using a conversion coefficient of each species that is provided by the ingested amount of radioactivity of the ICRP Publication 72 [13].

Although inputting cooking information is available and there are cooking information templates to be obtained from a recipe site to make the input easier, it is still a hard work to manage each meal every day. We want to consider the ways to make this process even easier. For example, we consider making it possible to identify food from a photograph of the cooking or the supermarket receipt and POS information of shopping to guess the food.

For determining the amount of the radioactivity ingested, the food eaten by the user might not be available in the food database, and does not correspond to any food measurement results; even if it is unknown it still may be sufficient. We do not consider the radiation level to be 0 but wait a month and build a graph of the uptake potential calculating it from the maximum value and the minimum value of radioactivity in food of the predetermined period on the food database width, or ingenuity, such as displaying the value of the radioactivity survey conducted before the accident [14]. In this regard we think it is necessary to pursue the reflection on the best way to proceed.

As mentioned earlier, the results for Cs-137 and other nuclide species are not currently integrated into the database structure to correspond to the multi-nuclide variety. This not only concerns nuclides which are listed in the radioactivity measurement results of the MHLW, but also the internal exposure control is possible by ingestion of natural radionuclides by incorporating models such as potassium 40 and polonium 210.

Also, due to the use of a computational model of the original CORPORE calculation system, used in the European context, we would like to consider the comparison method and use of the computational model with accordance to the Japanese data only.

Then, from the original CORPORE development team we receive an opinion that "since the genre of diet became different we want to be able to compare the exposure situations. Separating the diet patterns will help to make groups in the future automatically." For example, seafood diet group, agricultural group, etc. will help to build a kind of patterned templates for further study, and can increase the number of subgroups in each group such as according to gender or regions for future statistical data. In this paper, the system was evaluated by experts. The evaluation by residents is also required to evaluate the actual usability. We intend doing it in the future.

In addition, after the nuclear accident, there was an increasing interest in medical exposure; the efforts have become more active for the medical exposure dose management [15–17]. Also, Dose Structured Report was defined in DICOM as a standard for radiation exposure information management for medical exposure [18], while we are in the environment, which is easy to collect the information of the medical exposure.

As it is described above, in the present situation the influence of internal exposure from food is almost negligible for a large number of individuals although it is of concern for the residents and could be sensitive for specific groups of population consuming home-grown products and products from the forest. The impact of external exposure from the surrounding environment is considered to be potentially larger depending on the area as well as on the individual habits and activities. External exposure data have been gathered in the Prefecture health survey in collaboration with dose assessment [19]. In the future we consider building a system that can manage all the radiation exposures (including medical exposures) of the individual, with the main purpose to develop the radiation protection culture.

In addition, radiation dose information should not only be managed, but also the day-to-day nutrition management and health management could be considered also, with a unification mechanism where individuals can add nutrition information on food. Furthermore, it is expected to be also applicable for the management of the body accumulation of toxics such as arsenic and heavy metals and not only radioactivity.

20.5   Conclusion

Using the tools of the mechanism that has been used for internal exposure assessment in Europe after the Chernobyl nuclear power plant accident, we built a Web application for local residents of contaminated areas suffering from the Fukushima nuclear power plant accident so that they can determine themselves whether or not to eat the food, and use this tool to receive information. The user can input cuisine meals information in food ingredient units in accordance with the Japanese food environment. It is possible to view the quantity of radioactivity ingested and also the history of the internal exposure. It is designed to help to develop the radiation protection culture of residents by capturing the information visually.

Acknowledgement This work was partly funded by the 2012 fiscal year of Education, Culture, Sports, Science and Scientific Research. The results of this study are part of Fukushima Prefecture radiation medicine research and development business subsidy.

References

1. Lochard J et al (2009) ICRP Publication 111 – Application of the Commission's recommendations to the protection of people living in long-term contaminated areas after a nuclear accident or a radiation emergency. Ann ICRP 39(3):1–4, 7–62
2. Clement CH, Sasaki M (2012) The international commission on radiological protection and the accident at the Fukushima nuclear power stations. Jpn J Health Phys 47(3):206–209
3. Ministry of Health, Labor and Welfare (2011) Levels of radioactive contaminants in foods tested in respective prefectures. Available from: http://www.mhlw.go.jp/english/topics/2011eq/index_food_radioactive.html. Accessed Apr 24, 2015
4. Fukushima Prefecture, Monitoring Info (2011). Available from: http://www.new-fukushima.jp/monitoring/en/. Accessed 12 Jul 2015
5. Fukushima city, The result of independent radioactivity inspection of corps in Fukushima city (2011). Available from: http://www.city.fukushima.fukushima.jp/soshiki/22/enngeitokusann12073001.html. Accessed 12 Jul 2015
6. JA Shin Fukushima, Information of radioactivity monitoring (2011). http://www.shinfuku.jp/monita/. Accessed 12 Jul 2015
7. Hériard Dubreuil G et al (1999) Chernobyl post-accident management: the ETHOS project. Health Phys 77(4):361–372
8. Lochard J (2007) Rehabilitation of living conditions in territories contaminated by the Chernobyl accident: the ETHOS project. Health Phys 93(5):522–526
9. MySQL5.5 (2011) http://www.mysql.com/
10. PHPL5.5 (2013) http://php.net/
11. Apache2.4 (2012) http://httpd.apache.org/
12. Kawashima M et al (2015) Development and evaluation of a nutrition management system for elderly people with a dish registration function. IPSJ J 56(1):171–184

13. Japan Chemical Analysis Center (2010) Radioactivity survay data in Japan. 145(Aug 2010)
14. Ota T, Sanada T, Kashiwara Y, Morimoto T, Sato K (2009) Evaluation for committed effective dose due to dietary foods by the intake for Japanese adults. Jpn J Health Phys 44(1):80–88
15. Shannoun F (2015) Medical exposure assessment: the global approach of the United Nations scientific committee on the effects of atomic radiation. Radiat Prot Dosimetry pp. 1–4, doi: 10.1093/rpd/ncv027
16. Rehani MM, Frush DP (2011) Patient exposure tracking: the IAEA smart card project. Radiat Prot Dosim 147(1–2):314–316
17. Takei Y et al (2014) Summary of a survey on radiation exposure during pediatric computed tomography examinations in Japan, focusing on the computed tomography examination environment. Nihon Hoshasen Gijutsu Gakkai Zasshi 70(6):562–568
18. DICOM Standards Committee, WG (2005) Digital Imaging and Communications in Medicine (DICOM). Supplement 94: Diagnostic x-ray radiation dose reporting (dose SR)
19. Akahane K et al (2013) NIRS external dose estimation system for Fukushima residents after the Fukushima Dai-ichi NPP accident. Sci Rep 3:1670

Printed in the United States
By Bookmasters